Cognition of the Law

Luigi Cominelli

Cognition of the Law

Toward a Cognitive Sociology of Law and Behavior

Springer

Luigi Cominelli
Cesare Beccaria Department of Law
University of Milan
Milan, Italy

ISBN 978-3-030-07746-4 ISBN 978-3-319-89348-8 (eBook)
https://doi.org/10.1007/978-3-319-89348-8

Updated and expanded translation from the Italian language edition: Cognizione del diritto. Per una sociologia cognitiva dell'agire giuridico by Luigi Cominelli, © FrancoAngeli Editore 2015. All Rights Reserved.

Printed on acid-free paper

This Springer imprint is published by the registered company Springer International Publishing AG part of Springer Nature.
The registered company address is: Gewerbestrasse 11, 6330 Cham, Switzerland

Preface

Scarcely a few months had passed since this work was first published in Italian, and it was already becoming clear to me that the material was beginning to age. In discussions with colleagues, and in my follow-up research, a steady stream of stimuli flowed in that only amplified this awareness, and after 3 years, having drilled deeper into a range of specific issues, I was ready to give this work the refurbishment it needed. I proceeded to update the bibliography and enrich the text with numerous passages, clarifying my thinking and integrating it with perspectives and works that 3 years earlier I had not had time to consider. For this I thank all the colleagues and friends who offered their suggestions and shared their insights with me.

In the course of revising the book, I also realized that two specific subject areas had been covered a bit too hastily in the first edition, and they therefore deserved to be further developed. Thus, Sect. 3.3 has been extensively rewritten with a discussion of social epigenetics—which I believe, and hope, will increasingly become a field of interdisciplinary exchange among scholars in the humanities and other areas of study—and Chap. 4 now offers a deeper discussion of cognitive biases, as well as a new section (the seventh) devoted to the question of nudging. This expanded section is based on research I have done for an article that in the meantime has been published in the journal *Diritto e questioni pubbliche*, and I would like to thank its editors for giving me the opportunity to reuse that material, which fits perfectly into the scheme of this work.

Because the intersection of law, social science, and cognitive science is a topic of international interest, it felt appropriate to bring this work to an international audience, and here I would like to thank both my copyeditor, Filippo Valente, whose help in revising the text in its final English version is much appreciated, and the editor of the work's first (Italian) edition, Tommaso Gorni. Some of the passages in the text are quoted in translation: these translations are my own, and so responsibility for any error or misunderstanding that may have crept in rests solely with me.

Milan, Italy
April 2018

Luigi Cominelli

Introduction

This work finds its inception in a concern to understand what drives and motivates human behavior. As a constraining factor of behavior, the law can be used as a testing ground on which to validate sociological theory and form hypotheses that may integrate paradigms that have so far been irreconcilable. Even if we give up the positivistic pretense that sociology can be fashioned into an exact science, this does not relieve us of the task of solving its internal contradictions or its incompatibilities with other social sciences. In this work, I will accordingly outline a framework within which to reason about these issues in the social sciences, a framework in which sociology in a way acts as an overarching discipline. While physics has long been attempting a theory of everything—looking to reconcile general relativity with quantum theory—the social sciences have not been as interested in the analogous attempt to work the emotions and pulsions, including the sense of justice, into a coherent account of the way we behave under the prompts and restraints of the law. But there would be much to be gained by such an attempt, and this interest ought to accordingly be intensified.

This idea of reconciling the best paradigms originates and finds meaning with the inclusion of cognitive science in the sociologist's intellectual toolbox. In the cognitive sciences, those approaches that understand our biological nature to be inseparable from our social nature intersect, and I would submit that they yield the most promising hypotheses. One of the main sources of inspiration, initially more by happenstance than by design, has been my engagement with various issues in evolutionism. In school we make our first acquaintance with the basic principles of the life sciences, but then if we choose to pursue the social sciences or the law as areas of study, we may never cross paths with those principles again, or we will do so by chance or for the wrong reasons, when making a point of showing that they belong in another universe, and so that any mutual contact should be kept to a minimum.

During my postdoctoral research in Australia, I chanced upon Richard Dawkins's *Selfish Gene*: this was when I fully realized that the social sciences can only benefit from an updated and more advanced understanding of the natural world (with the usual caveat that we should steer clear of oversimple renderings of that knowledge). I would even speculate that the bleak beauty of the Australian continent provided the sensory spark for an insight I would go on to nurture, namely, that our natural history—now reflected in the palimpsest of that habitat and its captivating biological peculiarities—could not be entirely unrelated to our cultural history, which by comparison is much newer, only a few tens of thousands of years old. Something must have been wrong about the extreme culturalism that by inertia had become the standard paradigm in anthropology and sociology.

This was a suspicion I had already formed while studying conflict and the dynamics involved in disputes, a classic theme in the sociology of law and a through line that connects much of my research in this area of study. In investigating the interests that underlie a dispute, and observing that they sometimes lead to confrontation precisely when they coincide, it became apparent to me that we can only get so much mileage out of the standard paradigms that resolve interests into macro-variables. To this question I devoted the initial chapters of the book.

In the manner of a karstic river, never quite coming to surface, the intersection of the social sciences with the life sciences has long informed my more orthodox interest in mediation and dispute resolution. To appreciate this connection, we need only consider the use of law as a tool of conflict resolution, for this is a function that an evolved society develops to deal with the instinct for appropriation and the associated instinct for aggression. This in turn brings into focus the balance between the cooperative social instinct and the competitive individual instinct, a theme which clearly echoes the debate between communitarianism and liberalism, and which runs through all this work, and even today it engages the best social research.

Now that this work is complete, there is a sizable group of people I would like to thank for making it possible. Foremost among them are my beloved Silvia; my parents, Tiziana and Giovanni Sr.; and my entire extended family for the emotional, moral, and logistical support they have offered, so essential over these years, with the children still requiring much time, attention, and love: I hope that if one day Giovanni Jr., Francesca, and Rachele should find this book and leaf through its pages, they will understand all the time that was spent away from them. It is in the first place to the three of them that my effort has been devoted, and as much as I did want to leave something behind that can be appreciated for its content, this intellectual ambition I cannot separate from that first commitment.

It was mentioned at the outset that this content tends to age quickly, because research in these fields is advancing at a rapid pace, putting forth a constantly evolving body of knowledge. Even so, I like to think that at least some of the ideas and arguments presented here offer an original contribution and will have some lasting power, continuing in the future to provide useful insights into the social arrangements and institutions that form the background to the natural history of humankind. One person's ambition is, of course, nothing by comparison with the size and scope of that process, with its intersecting planes of nature, society, and history, but if this work manages to capture even only an infinitesimal part of it, or points us in the right direction in that regard, I should count myself happy.

I also owe a lot to all my mentors and colleagues for engaging me on the problems tackled in this book: their intellectual stimuli have breathed energy into this project. In this group I include Vincenzo Ferrari, Masayuki Murayama, Alberto

Toffoletto, Rick Mohr, Giuseppe de Palo, Valerio Pocar, the late Morris Ghezzi, Edoardo Fittipaldi, Maria Cristina Reale, and Letizia Mancini: at different times and in different ways, wittingly or otherwise, they have each had a role for which I am very grateful. With each exchange, they have made it possible to peer a little deeper into the mystery of humanity, and even if by dint of study this mystery may become every day less of a mystery, it cannot cease to enthrall the mind: we cannot turn our gaze away, and I hope to preserve that sense of wonderment going forward.

Milan
December 2015

Contents

Chapter 1
The Social Subject as a Legal Actor

1.1 The Social Subject as a Legal Actor

Sociologists who deal with legal phenomena often live a paradox and an internal contradiction that they are not always able to admit or to identify or resolve. This is especially true of holistic approaches to socio-legal studies, for which the individual is in some way, at different levels of depth, a function of the social group or the social system. The legal phenomenon is in fact intrinsically individual. Every modern legal conception, wittingly or not, is based on individual rights and duties. The collective implications of the law are perfectly clear to lawyers and laymen, and when needed they can be translated into institutions such as collective responsibility or group sanctions, but the common unit of legal analysis is the individual with some legal capacity. Think of the Roman brocard *Haec sunt principia iuris: honeste vivere, alterum non laedere, suum cuique tribuere* ("The basic principles of right are: to live honorably, not to harm any other person, to render each his own" [Digest 1.1.10; Watson 1985, p. 2]). The very fact that we are now witnessing a renewed debate on collective rights and action shows that the perception of the law we have inherited has always been intrinsically individualistic.

On the purely holistic conception, the law is battened and flattened down to a dimension that is not the one through which it is perceived. There would be nothing wrong with this if the holistic premise had been consistently developed, but holistic sociology does not, in its analysis, appear to have thoroughly explained the individual social tendencies that are closely related to our feelings and attitudes about law and justice. For example, it has failed to explain a generalized tendency to comply with rules, to cooperate, to seek a solidarity of interests and to participate in solidaristic schemes, even allowing for all the distinctions that need to be made in accounting for all those areas of social life where anomie prevails. And when the holistic conception *has* explained these attitudes, it has done so using ethical or political paradigms it has not been able to then rationalize according to the canons of a modern falsificationist social science.

L. Cominelli, *Cognition of the Law*,
https://doi.org/10.1007/978-3-319-89348-8_1

What I am going to do in this book is to outline in the first place the methodological assumptions needed to analyze *personal legal cognition*, understood as the complex of activities by which we each perceive and analyze the law and make decisions in connection with it. This work will make use of expressions such as *legal attitude*, *emotion*, and *feeling of justice*. It would be vain, however, to define such concepts here in a way that satisfies all validity tests in the different disciplines that will be considered. From time to time I will provide stipulative definitions, not necessarily validated on an empirical basis. The term *attitude* will be brought to bear on the law by borrowing it from social psychology, and so from a discipline that constitutes an acceptable link between sociology and the behavioral sciences, but without thereby having to proceed from the same epistemological premises as this science. Emotions and feelings will be analyzed by drawing not only on psychology but also on the partly differing definitions given by sociology.

The cognition of the law will be explored from the perspectives of sociology, social psychology, evolutionism, sociobiology, and the cognitive sciences. The cognitive sciences, in the field of human behavior, are today proposing a synthesis, with a welcome reversal of the compartmentalization of scientific investigation. In this exploration, I invoke a "weak physicalism" (Meyering 2000) that will be better expounded in Sect. 1.5. In essence, this conception deems it feasible, within the epistemological autonomy of each science, to connect the different fields of knowledge through "bridge-laws." The clarifying operation I have in mind is explicitly neopositivist in its inspiration, with an intent to explain human affairs, and I will try as far as possible to maintain this intent separate from a consideration of practical-ideological and aesthetic aspects—an investigation that is often useful in the social sciences, and whose full membership in the field of scientific research is undoubted. The confusion between these aspects, however, has so far made sociology a science that is problematic and perceived as a purely interpretive discipline, closer to literature or to political discourse than to the canons of science (Collins 1975). Contemporary socio-legal studies have not been exempt for this problem of interpretivism (Tamanaha 1997, p. 253).

In order for the law to be considered at least partly independent in the effort to explain the legal attitude, it is necessary to refer to some updated interpretive principles of "methodological individualism" (Boudon 1987), so as to overcome the objections of having surrendered to behaviorist reductionism. This leads us directly to the importance of distinguishing between macro- and micro-approaches in sociology (Alexander et al. 1987). In fact, according to Jeffrey Alexander, the epistemological dispute over reductionism has been replaced by a search for the link between the macro and the micro (Alexander and Giesen 1987, p. 3). The unresolved relationship between the micro and the macro may be no less than an underlying cause of a supposed crisis in sociology (Addario 1994b, p. 7), as if they were incompatible paradigms whose coexistence could discredit the field. In this, we can view the simplistic contrast between the individualism of Anglo-Saxon liberalism, which produces various micro-theories, and the communitarianism of continental European social-democratic progressivism, rooted in German idealism and French revolutionary naturalism.

The most productive arguments for a purely macro-theory in sociology derives from Marx, who argues that individual motives are irrelevant in sociological analysis (Alexander and Giesen 1987, p. 6). The absolute prevalence of the macro comes through in Althusser's structuralism and in Luhmann's functional structuralism: extreme constructivism postulates objective social structures that overshadow subjective ones, asserting that the actors' subjective inclinations cannot be considered independent variables. Among the micro-theories that have had a major impact on sociological theorizing is Mead's (1972) pragmatic philosophy, which translates into sociological theory by likening society to a game.

The psychoanalytic theory of society denied that the group or the collectivity had any necessary function, and linked them instead to individual fantasies or pathologies that could be corrected in a process of rationalization. The group was only a gathering of individuals, and it had to be defined as such in order not to fall into the fallacy of a "group mind" or of a "gigantic Moral Being" (Malinowski 1939, p. 938). Even the social sciences in the end forgot that institutions are functional to society only because they are functional for individuals (Homans 1961, p. 384), and that it was necessary to bring individuals back into the analysis, because only real and physical people can make social phenomena happen. Behaviorism, on the other hand, is faulted for depicting action as a stimulus-response sequence, and depicting learning as a complex of material experiences agglomerated by way of physical reinforcement.

The historical contingency that sociology derives intellectually from philosophical reasoning brings even further complication to the attempt to reconstruct the nexus between micro- and macro-analysis. The synthesis between economic and behavioral individualism, psychoanalytic individualism, Marxist materialist collectivism, and Durkheimian collectivism is, finally, well represented by Weber and the ideal-types of social action, according to which society is a collective force created by individual and contingent acts of freedom (Alexander and Giesen 1987, pp. 14–15). The focus falls not on individual action as such but on the typical *motives* of action. Here Weber's contribution is disruptive: the recognition of affection, charisma, and leadership as means of social action leads to the final recognition of the micro level. And yet Weberian analysis is undeniably methodologically closer to Marxian analysis, because it presupposes action in a material and instrumental manner, and there is not much room left for individual motivation and inclinations. Next to interpretive sociology, Weber develops a macro-structural sociology of a comparative and historical kind (Addario 1994b, p. 26).

Schütz acutely points out that Weber does not distinguish acting as an "ongoing action" from acting as an "accomplished act" (even though this is irrelevant for the purposes of Weber's analysis), nor does he distinguish the meaning of action attributed to it by the actor from the one attributed by others (Schütz 1967, p. 30). Weber, in other words, does not distinguish between self-observation and external observation of the act, making nonrational or affective behavior into a limit case of meaningful behavior in general, even though affectual behavior can be subjectively meaningful. Affectual and traditional behaviors are adopted by those who satisfy their own need for revenge, enjoyment, dedication, and respect for tradition through an immediate release of emotions (Weber 1978, p. 24). In Weber, the definition of action as a meaningful behavior is not ultimately relevant for an analysis focused on the individual. When it comes to meaningful behavior, Weber always has instrumental rationality in mind. The classification of ideal types of action is carried out by equating the meaning attributed to an act with its cause, and this confusion leads to a logical fallacy.

Moreover, the ultimate aim of interpretive sociology is to outline an *objective* interpretivism. The perspective of individual social cognition is not functional to Weberian discourse. This clearly emerges in the references that Weber makes to the contents of the social relationship, such as "conflict, hostility, sexual attraction, friendship, loyalty, or economic exchange," or "the fulfillment, the evasion, or the violation of the terms of an agreement" (Weber 1978, pp. 26–27). These factors are never considered from a subjective point of view but are rather objectified to form the interpretive framework of social action. In fact Habermas (1986, p. 231) concludes that the fundamental concepts of action formulated not only by Weber but also by Marx, Horkheimer, and Adorno are not complex enough to give us an understanding of how social rationalization can be applied to social action.

Also belonging to the micro- and individualistic tradition are ethnomethodology and symbolic interaction, which propose a sociology of action in everyday life (Garfinkel 1967; Goffman 1959). In this case, each macro- and superstructural theoretical apparatus is deconstructed with an awareness that any superior meaning conferred on action is illusory. A status, a position—in short a social niche—is not something material that can be possessed and then exhibited (Goffman 1983, p. 217): the patterns of appropriate behavior are something that must be put on stage, illustrated, and realized, in a deliberate way with either cunning or sincerity. For ethnomethodology, the social order rests solely on factors that are largely unconscious and not entirely deliberate, and which for the most part pass unnoticed, whether by the actor or externally. Sociologists have the common tendency to reduce human societies into nonacting social unities, such as the social class (Blumer 1969). Society is, in the first place, social action, and human action is not "a sheer release of an already organized tendency" or a tendency "sweeping the individual along in a relentless move towards its realization" (ibid., p. 96), but is rather a "construction built up by the actor" (ibid., p. 94). There is talk of a new "action language," where every static social situation needs to be translated into corresponding social action (Emirbayer 1997, p. 307).

In this way sociology moves closer to ethology, which Goffman draws on in his research. The methodological tools used are the analysis of *situated activities*, the reconstruction of the meaning of expressions *given off*, and the *reflexivity* through which individuals explain and rationalize their everyday experiences. The relativization of each macro-explanation is so deeply rooted that the dramaturgical analogy proposed by symbolic interactionists—the paradigm around which their sociological explanations revolve—is recognized to be, at least in part, the result of a theoretical stratagem, and something not to be taken too seriously, even though it is intended to describe the social structure of those entities of social life that exist whenever people come within the immediate physical presence of others (Goffman 1959, p. 160).

The ethnographic approach contradicts the normative order of sociology and draws sociology into an interpretative and situationalist sphere. Descriptive procedures, their results, and the use of these results are elements integrated into the very social order these procedures mean to describe (Cortellazzi and Piccoli 1988, pp. 109–111). Social order thus becomes a representation of individualities. It has been argued (Bovone 1994, p. 55) that Goffman's modernity and longevity were due to an excessive attention to appearance and its theoretical justification—a distinguishing feature of our era—and that ethnomethodology has not properly worked together philosophical and sociological concepts. Others (Ciacci 1983, p. 15) have criticized Goffman as overly reliant on participant observation with easily manipulated underdogs. Critics also charge that interpretive and hermeneutic microsociologies are bound to fall into contradiction and to severely undercut their individualistic stance: their emphasis is not on subjectivity as such but on interpretive activity and its rules (Addario 1994b, pp. 15–16).

It seems to me that these critics are not doing justice to these early, sometimes provocative attempts to uncompromisingly adapt sociological theory to field research and observation in the proper sense. Even if the explanatory outcomes have not always been satisfactory, these microsociologies have deservedly challenged the sociologist's faith in a perfect system of rationality, a faith that had wound up distancing sociology from social phenomena. They questioned the systems of social norms conceived without analyzing face-to-face interaction and without taking the "irrational" factors into account. In Sect. 1.3, I will devote further analysis to the concept of rationality, arguing that what has been discovered so far is largely inadequate.

The great merit of ethnomethodology has been to attune researchers to the need to take the situational variables of every social situation also into account: actors as objects, the scenario as such, evoked meanings, the time occupied by interaction (Denzin 1983, p. 230). I will try to show that this is one of the most important crevices through which to inject sociological dogmatism with the skepticism needed to build a phenomenological foundation for an *integrated social science*. The symbolic perspective of interactionism and ethnomethodology, however, is not the only theoretical paradigm to have seeped in, claiming to have overcome positivist reductionism through disciplinary crossovers, and explaining "ordinary, everyday social behavior" (Homans 1961, p. 2). Thus, unless sociologists want to confine themselves to work of a purely descriptive sort, they would do well to look to psychology, for example, and borrow the nomological propositions it has to offer concerning collective phenomena (Williams 2001, p. 136). In Homans's view, primal social groups come into existence as methods of mediation between the individual and natural environments (Sims Bainbridge 1997, p. 92). In time, the evolving group brings other purposes into focus, against the background of a cosmological scenario that alternates catastrophe, stasis, and navigation. Human behavior is a function of the resultant reward or punishment, and it becomes social when it triggers a similarly determined behavior in another individual (Homans 1961, p. 13).

Homan's proposal for facilitating the coexistence between the analytical primacy of society and the explanatory primacy of the individual exceeds the perspective of the *homo sociologicus*, an agent entirely determined and produced by society. The attempt to explain everyday social behavior, combining psychology and economics, taking the relative special conditions also into account, proceeds from the shared aspiration to overcome conceptual schemes that are logically unassailable but devoid of explanatory value (ibid., pp. 10–12). The theoretical results that Homans achieved seem somewhat idle to us today in light of behavioral economics and the cognitive sciences, as they ultimately reduce social patterns to four fundamental types (influence, conformity, competition, and esteem). He did, however, make very insightful analyses, as when he commented that while game theory may provide "good advice," it is also a poor description of human behavior (ibid., p. 81).

Just at the same time, sociological behaviorism was fully expressed in Mead's work (Mead 1972). This was a cultural and social behaviorism, not an innate and individualistic one. Mead attempted to show that the mind and the self are social emergences without residues, and that mechanism for their emergence is that of language in the form of vocal gestures. Here too, the approach is strictly individualistic: "individual experience and behavior is, of course, physiologically basic to social experience and behavior" (ibid., p. 2). The process of social differentiation has two poles: one individual or physiological, the other institutional (ibid., pp. 229–230). How to explain the individual's social behavior? The social act involves cooperation among several subjects. The act is preceded by a gesture whose consequences the author cannot foresee except by developing a symbolization capacity that allows him to evoke and assume the point of view of the *alter ego* (Ciacci 1983, p. 27). By so anticipating the concept of empathy, the mind's functioning processes are situated in the ability to anticipate reactions and to step into the role of a *neighbor*, ultimately modifying a chosen course of action (ibid., p. 31).

Mead's synthesis of psychology and sociology leads to the identification of some fundamental socio-physiological impulses or behavioral tendencies that break down into two classes: those that lead to social cooperation and those that lead to social antagonism. Even if the former are of a friendly nature and the latter hostile, this does not mean that antagonistic impulses of the second type are in some sense asocial. They are society-wide impulses (such as self-preservation and protection): they cut across the whole of society and are controlled and restrained by the state's legal system (Mead 1972, p. 305). Mead did not manage to achieve his ambitious goal of constructing an "empirically based psychological theory that would make it possible to ground the genesis of human reflective behavior in the physiological dimension without reducing it to the latter" (Baggio 2015, p. 15). But I think this attempt has disproved Durkheim's apodictic assertion that every time a social phenomenon is directly explained through a psychological phenomenon, we can be sure that the explanation is false (Durkheim 1917, p. 129).

Symbolic interaction, ethnomethodology, and sociological behaviorism give evasive answers to problems of intersubjectivity, that had until then been formulated by way of overly complex questions. In a way, these approaches came in response to a wide demand for applied sociology, and to the dissatisfaction with the answers of sociological functionalism. Cross-contamination between disciplines, and an appeal to the contributions of philosophers and psychologists, echoed Weberian interpretive sociology (Ciacci 1983, p. 12). But there was not yet a clear call for a common methodological framework, or for the unified science of behavior that in the social sciences is still explicitly rejected.

1.2 The Connection Between Micro- and Macro-Theories

Micro-sociological and reductionist theories have culminated in the recovery of individual action and social behavior. Methodological individualism holds that sociological terms of reference must be translated into behaviors or observable traits of social actors: "On the basis of universal observational language, sociological terms could be translated into psychological terms and on that basis a unified science of human behavior would emerge" (Alexander and Giesen 1987, p. 20). This problem makes it necessary to more accurately set out the relation between the micro and macro spheres and to analyze some fundamental paradigms by which this relation is defined.

The dispute between the macro and the micro perspective has been defined as essentially ideological and not properly methodological (Addario 1994b, p. 12). Its origin may be located in the "divergence between spiritualism and materialism" (ibid., p. 18). In a context in which priority is given to the individual, both analytically and theoretically, theories fail to integrate the individual and social dimensions. But when priority is accorded exclusively to the social, I agree with the assertion that theories cannot explain *anything* (Williams 2001, p. 149). The analysis of collective phenomena should consist of two steps: a social explanation of human behavior (the bridge-problem) and an explanation of how social behavior is transformed into a collective phenomenon (the transformation problem) (Wippler and Lindenberg 1987, p. 149). Still, the most relevant analytical distinction should not be between the individual and the collective but between action and structure (Knorr-Cetina 1981, p. 17).

One of the most convincing approaches in this sense is offered by Raymond Boudon (2009). While it is true that, from a logical point of view, the distinction between the micro and the macro does not necessarily correspond to the distinction between individual interpretation and structural interpretation, it is also true that their overlap is virtually automatic. Boudon attempts to integrate the definition of the micro and macro fields by drawing on the distinction between interactionist and deterministic paradigms. Collins defines interaction as the key to microsociology. Interaction is where the energy of movement and change, the glue of solidarity, and the conservatism of stasis reside (Collins 2004, p. 3). The interactionist paradigms described by Boudon cyclically alternate, corresponding to methodologies of interpretation that are historically predominant in a given period.

The first interactionist paradigm identified by Boudon is Marxian. Marx's theory is so multifaceted that, depending on the perspective from which it is considered, it would be possible to also include it among those paradigms where priority is given to the individual. Unlike others, Boudon emphasizes Marx's perception of a history that is still the result of human activity. Indeed, the actions of individuals are independent and indeterminate even if individual preferences are predetermined and systemic (Boudon 2009, pp. 195–196). A second interactionist paradigm is identified by Boudon in Tocqueville's thought: here the actions of individuals constitute independent variables and are placed in a state of nature in which preferences are not given, and these, too, need to be explained individually. The third interactionist paradigm is Merton's: here, the assumption is one of free individual actions, and the interaction does not unfold in a state of freedom but rather takes place within the limits of a social contract. The fourth fundamental paradigm is, finally, the Weberian model, where intentional action is set within a framework of elements, both anterior and assumed, that reflects not only the prevailing system of preferences but also the associated choice of means (Boudon 2009, pp. 197–199). We should add to this a fifth interactionist paradigm, that of the "interaction ritual chains" (Collins 2004):

> The central mechanism of interaction ritual theory is that occasions that combine a high degree of mutual focus of attention, that is, a high degree of intersubjectivity, together with a high degree of emotional entrainment—through bodily synchronization, mutual stimulation/arousal of participants' nervous system—result in feelings of membership that are attached to cognitive symbols; and result also in the emotional energy of individual participants, giving them feelings of confidence, enthusiasm, and desire or action in what they consider a morally proper path. (Collins 2004, p. 42)

Deterministic paradigms interpret the behavior of the social subject as determined only by the elements that precede it. These paradigms, according to Boudon, are only of descriptive import. Their poor scientific value means that they can be pejoratively described as *sociologisms*.

The paradigms that most neatly fall into the deterministic mold are the *hyperfunctionalist* ones of *Mertonian* derivation: they are over-simplifications that Merton himself would have rejected, and they relate to structures of interaction set in a contractual context. The starting axiom is that (a) no action takes place in the context of a state of nature, (b) sets of roles and statuses are made up of complementary and noncontradictory elements, and (c) their range of interpretation is in any event inexistent or negligible. In a second version, which Boudon has described as *hyperfunctionalism of Weberian derivation*, the reductivist simplification traces every action to socialization effects. If preferences are imposed on the individual by social structure, it is then possible to identify a paradigm of *totalitarian realism*. Finally, Boudon singles out a paradigm of *methodological determinism*. This is the only scientifically acceptable paradigm, as it tries to explain macro-variables by reference to other macro-variables, claiming, for example, that a person's level of education explains his or her socio-professional level (Boudon 2009, pp. 239–242).

Boudon expressly criticizes the hyperfunctional paradigms, most typically associated with Bourdieu: the concept of *habitus* explains the fact that social classes act, reproduce, express themselves through individuals. As a result, reality is axiomatically irrelevant, and the individual cannot be considered a real creator of its meaning

(Crespi 1989, pp. 49–50). Boudon clearly reveals how this translates into a tautological paradigm that has poor explanatory potential, if not by virtue of post factum reconstruction. The existence of social regularity does not imply that social behaviors are directly deducible or produced by social structures. Several issues are raised by Boudon even as regards methodological determinism. While methodologically acceptable, this form of determinism falls into the reductionist fallacy, and it loses track of subjectivity and individual variability. According to Boudon (1969, p. 268), the impossibility of working by induction from collective to individual correlation has been sufficiently demonstrated without taking the extreme stance that the environment has no bearing.

A classic example of the individualistic paradigm is Tocqueville's famous explanation of the differences between French society and British society, and in particular his explanation of the macro-variable consisting of the development of agriculture as a function of the aggregate behaviors of landlords (de Tocqueville 1840, p. 153). Tocqueville employs an implicit theory of rationality that is far more general than economic theory. From a sociological perspective, the aggregation of monetary units is only part of the problem, because many individual and collective decisions concern nonmarket social exchanges, which cannot be monetized (Gerstein 1987, p. 92).

Aside from having been faulted for violating the unimpeachable principles of the sociological tradition, the individualistic paradigm has also been claimed to be useful only in analyzing Western individualistic societies. However, although community ties are supposed to be stronger in traditional non-Western societies, this does not mean that their social dynamics cannot be analyzed as a result of individual behaviors (Boudon 1987, p. 56). Nothing should rule out the use of individualistic paradigms in sociology. As we have seen, these paradigms have the advantage not only of expressing general laws, but also of affording ad hoc individual explanations of the singularities of macro-phenomena and of the differences between social systems. The use of individualistic paradigms entails greater theoretical challenges than the use of holistic paradigms.

Though methodologically neutral, the micro/macro distinction, as well as the distinction between individual and structure, easily turns into an ideological dispute. Sociological constructivism by and large accepts that descending to the individual level may work for organizations or small groups, but it does not accept that methodological individualism can serve in macro-studies: it considers this an improper imitation of economics. In my view, this latter critique is a form of sociologism, that is, an attempt to explain the social solely through the social, in a form of sociological degeneration that styles itself as "a true anthropology of itself" (Donati 1983, p. 61). No less deterministic is the attempt to relate every content or fact of social consciousness to some social event.

In this work I speculate that the contradiction must be resolved through an integrated vision that starts out from the idea of the subject as a "social individual." I also argue that the ontological distinctions between the *individual* and the *social structure* are incorrect when they insist on conferring methodological preeminence on one or the other. A sociology that is aware of the other behavioral sciences has all the tools needed to avoid behavioral reductionism, as well as the "sociological creationism" that we find in deterministic paradigms. To this end, I have identified at least two ways in which the gap between the micro and the macro has been convincingly closed.

The first is introduced by Jeffrey Alexander (1987, p. 289) as a "microempirical model of individual action, according to which action is viewed as interpretation and strategization." Alexander argues that "macro-theorists," by only recognizing actions presupposed in their narrow conception of collective order, have erroneously reconstructed the whole from a single variable. In other disciplinary fields, economic theories have argued that action is strategic, that cultural theories have accentuated its typification, and that theories of social movements have characterized inventive and creative action as fundamental (ibid., p. 314). Instead of being framed as dependent or independent variables, social structure and individual action may be described as "parameters and variables in an interactive system comprising different levels of different size," within a common conceptual framework in the social sciences (ibid., p. 315).

A wealth of illustrations of how the difference between the micro and the macro can be relativized comes from the history of medical science. Physiology at first defined its field in terms of the relation between the organs and the blood. Biochemistry later studied the changes that take place in the chemical properties of the elements the organs are made of. Then molecular biology interpreted the fundamental dynamics of life as molecular and atomic interactions. The nodal point is now to determine at what level it is more appropriate to study a particular phenomenon. In fact, Simmel claimed that we need to draw as clear a distinction as possible between real beings, understood as "objective" units, and their recomposition into complex units, which as such exist only in our "synthetic" spirit: the issue of how many and which real units we must synthesize into a higher unit is only a matter of praxis (Simmel 1982, pp. 14, 16). The micro/macro distinction is therefore relative, and the different levels should not be understood to be in competition (Alexander 1987, pp. 290–291).

The second convincing attempt to close the micro/macro gap comes from Collins (1987, p. 196), who contends that the macro and micro levels should not be treated as making up a *dichotomy* but should rather be distinguished along a *continuum*. And if macro-sociology should realize that its work is theoretically incomplete, radical micro-sociology must avoid excessive idealism by placing cognition within the experience of real people (Collins 1981, pp. 91, 94). Only time, space, and the number of combinations of micro-situations constitute valid macro-variables, and these cannot be translated into micro-situations related to personal experience (ibid., p. 81). When asked what props up the social structure of the macro-world, Collins replies: *property*, that is, the individual's relation to the physical objects that make up his or her world (ibid., p. 103). The most remarkable aspect of micro-behavior lies in the *enactment of property*, which amounts to nothing more than the appropriation of a particular physical space (ibid., p. 205). This would provide a theoretical reply from the standpoint of a "microtranslated macrotheory" and not a "reified macrotheory" (ibid., p. 204). In these passages we find an anticipation of the cognitive theories on the "endowment effect" bias, which will be discussed in Sect. 4.5. But suffice it to note here that realistic and phenomenological sociology has in this way found an empirical anchor through which to solve a problem which in some respects may be properly considered a false problem, but which, on the other hand, has continued to offer an excuse and a theoretical pretext to build self-referential and reified "closed" systems.

Finally, it is also worth mentioning that there have been several original attempts to reread Parsons's structural-functionalist theory as a well-conceived, if still incomplete, attempt to overcome the dualism between the macro and the micro (Addario 1994a, p. 81). In particular, Parsons pointed out that the distinction between the macro and the micro had no real substance: it rather hinged on a linguistic choice and hence on a theoretical stipulation. Parsons (1964) used the Freudian Superego to tie the micro and the macro level in sociological analysis, although he believed that Freudians were overly concerned with the effects of individual instincts and the negative effects of frustrations, and so he found their psychology too "biologically oriented," at the expense of the social and cultural factor (Kurzweil 1987, pp. 242–244).

Parsons's proposal was essentially to reestablish a connection between the Ego and the Superego and to emphasize their shared cultural aspects. In order to continue along the path started by Parsons—who neither wanted to, nor could, drive this epistemological revolution to its extreme consequences—as well as to avoid a "sociogenetic perspective" intended to explain the emergence of the social from the biological or presocial state, Addario (1994b, p. 14) proposes adopting a "constructivist and antirealist conventionalism" (Addario 1994a, p. 83) and introducing an "emergent effect" concept. It seems to me, however, that this does not really put an end to dualism: it merely updates the deterministic paradigm.

1.3 Rationality

The first of all the variables that influence legal attitudes, and indeed attitudes toward all kinds of normative phenomena, lies in the dichotomy between reason and emotion. In Weber's thought, modern legal orders are the highest point of rationality, understood as the deliberate calculation of one's actions in light of the goal to be achieved. The use of the rational-irrational and formal-material variables enables Weber to typify the historical forms of law (Weber 1978, pp. 654–658, 809ff.). Weber's analysis, however, does not address individual attitudes toward law, and only in outline, or at a level of historical-cultural reconstruction, does it explain the attitudes and cognition of the affective actor and the traditional actor. In Weber's work, the traditional actor, in particular, reacts in an automatic and perfunctory way, through acquired habits that conform to daily life (Fitzi 2011, p. 39).

Pareto's analysis revolves around the logical/nonlogical variable and focuses on numerous individual attitudes that today could form the background to, and basis of, a cognitive sociology of legal action, contributing to its theoretical foundation.

Logical actions are, for Pareto (1981a, p. 81), "means appropriate to an end," while nonlogical actions originate from a psychic state (feeling, subconsciousness) that has nonverbal manifestations. Pareto seems to follow the Cartesian *summa divisio*, believing that our actions are explained by a constant, nonlogical, and instinctual side, and by a deductive, logical side aiming to explain and justify the first. It should be pointed out that Pareto does not consistently follow his own conceptual scheme, because he seems to forget it and freely manipulate it when he needs to explain real facts. The drivers of action, as is well known, are *derivations* and *residues*. Derivations are the logical-conceptual rationalization of residues. The most relevant derivations for the law (Pareto 1981c) are *assertions of fact*; *authority*; and *what is in accord with individual interest, collective interest, or juridical entities*. The residues (Pareto 1981b)—the nonlogical sentiments that form the permanent substance of every human decision-making option—are divided into six classes: (1) residues of combinations, (2) residues of the persistence of aggregates, (3) residues of the manifestation of sentiments through external acts, (4) residues of sociability, (5) residues of the integrity of personality, and (6) residues of sexuality. The six classes of residues are variously redefined at different places in Pareto's work.

The *persistence of aggregates* describes the human tendency to conform to habits or individual patterns of behavior (Pareto 1981b, p. 19), which clearly conjures up "traditional" social action. The persistence of aggregates is therefore the observable tendency in human behavior to consolidate sets of feelings, ideas, and ways of acting into permanent units over time. The instinct of *sociability* outlines the observable tendency toward cooperation and uniformity in treatment (ibid., p. 162). The instinct of *integrity* describes the tendency to react to offenses entailing exclusion or marginalization by the community (ibid., pp. 293–296). Residues often wind up overlapping and creating "compound residues," all of which form a sense of justice (Simonelli 2006, p. 66). The Paretian theory delineates an explicitly realistic and conflictual vision of the law. It could also be considered one of the first fully developed cognitive sociologies of law, because it explicitly captures the psychological-behavioral aspects underlying each individual legal attitude and the laws explaining the development of society (Pareto 1980, pp. 580–581).

Pareto's belief that both economic and social phenomena should be considered a result of individual actions lays the groundwork for an integration between the behavioral sciences. It also emerges that there is no real structural distinction between derivations and residues or between rationalization and emotion. In fact, derivations are a post factum rationalization of an already made choice and an already taken decision, and hence a superstructure relative to residues, which are the great irrational forces that drive history (Simonelli 2006, p. 70) and determine nonlogical actions, that is, the most numerous and determinative of historical-social events.

Pareto's view is already on the path toward a modern cognitive conception. Humans tend to reassure themselves by hiding their feelings, in order to rationalize and mask the motives of action with logic. The rationalization of one's feelings brings a competitive advantage if competition is correctly understood as "the fight of all *for* all" (Simmel 1955, p. 62) rather than as a purely zero-sum game.

Rationalization, however, takes place unknowingly, mostly in good faith. This kind of rationalization closely resembles those processes of self-persuasion which will later be defined as the cognitive bias of self-confirmation, and which we now understand to be an adaptive trait that has evolved to give us a social edge.

Among the *metaphysical derivations* impacting law, Pareto identifies solidarity, progress, humanity, and democracy. These are abstract entities entrusted with the pursuit of the common good. But, once the superstructure that provides an ideological justification is removed, they are just names that indicate indistinct and incoherent feelings (Pareto 1981c, p. 83). The original substance of positive law consists of residues (*law as fact*) that only subsequently are structured by legislation, legal scholarship, or the case-law (*law as theory*). Judicial decision also depends in large part on personal inclinations, shaped by such factors as political influence, ideological commitments, and even temporary impressions or pure luck.

Pareto's point of arrival is a theory of legal ideology that is realistic and resonates with contemporary conceptions of *living law*, legal realism, and sociological jurisprudence. This is associated with a political concept of law that is integrated with a consistently anti-utopian conception of élites. A government, whether despotic or democratic, cannot impose measures deviating from the residues that are prevalent among its subjects, or from common sense. Residues also become something material and distributed to different extents among social classes: this in a way legitimizes inequality, because on the basis of a different capital of cognitive skills, the élites will have more such skills.

Even the reconstruction that Merton offers of individual modes of adaptation is based on a rationality intrinsic to action. Individuals act on the basis of a personal drive that can be inspired by conformity, innovation, ritualism, retreatism, or rebellion (Merton 1968, p. 194ff.). Merton typifies a number of behaviors: he does so in the first place by referring to individual attitudes toward the importance a social objective has in the culture in which the individual is embedded, and toward the norms and channels that society points to as institutional and therefore legitimate. Merton mentions and anticipates the themes of rational choice and bounded rationality by introducing, for example, the "rational" criminal (Merton 1968, p. 244ff.): one who acts rebelliously when faced with a situation in which the expectations of individual success are high but the means by which to achieve that success are scarce.

In the *ritualist*, Merton sees coerced conformism, which takes foot in particular in bureaucratic organizations, along with corporatism and deresponsabilization. The *conformist* responds to threatening situations with a "prophylactic avoidance" of the approaching situation: he is not aware of being afraid, and he is satisfied with, and proud of, himself because of his foresight (Merton 1968, p. 240). Through the conformist attitude, an act that originally had *instrumental* significance is transformed into an act that is *inherently* valuable, with no further objective, and this becomes the distinctive mark of a neophobic society. Ritualism is strongly linked to authoritarianism, on which Adorno's research group had worked a few years earlier (Adorno et al. 1950). In Merton's typification, the ritualism of compulsive conformists entails a series of elements as follows: a strong preference for symmetry, that is, for what is familiar, definite, and regular; a tendency to prefer uncompromising solutions; overly simplified dichotomies; absolute either-or choices; rigidly determined options; hasty and preconceived conclusions; obstinacy and stereotyped solutions; an inclination to revere formalities, through a diffuse *globality*, or by overplaying the importance of a particular detail, or through a compartmentalization of *stimuli*; and ultimately an intolerance of ambiguity (Merton 1968, p. 241).

As with the social ritualist, Merton attributes choices to the individual that are dictated by "syndromes": in this case, a love for safety and a rejection of risk. In the case of the *retreatist*, Merton explicitly mentions "the adaptive activities of psychotics, autists, pariahs, outcasts, vagrants, vagabonds, tramps, chronic drunkards and drug addicts" (ibid., p. 207). He notes that sociologists have been uninterested in these forms of deviant behavior, despite the social antecedents and the manifestly social consequences, and this has cleared the way for psychiatrists, who reframed these attitudes under the rubric of apathy or melancholy (ibid., p. 243).

Here we need to go back to the analysis of authoritarianism and authoritarian personality, for it shows that individuals adaptively realize that they can benefit from unconditionally obeying the orders of authority (Ravenna 2012, p. 156). For a long time, these concepts were the main reference points for theories of nonlogical or irrational action. Irrationalism is cited among the ideological causes of twentieth-century totalitarianism: "The goal toward which the pseudoconservative mentality strives—diffusedly and semiconsciously—is to establish a dictatorship of the economically strongest group. This is to be achieved by means of a mass movement, one which promises security and privileges to the so-called 'little man'" (Adorno et al. 1950, p. 685). By educating individuals to obey distancing them from the victim's suffering, totalitarian regimes found easy success. The key to explaining the aptitude for orderly and obedient observance of the established order was an authoritarian syndrome that could be understood psychoanalytically as a sadomasochistic solution of the Oedipus complex (ibid., p. 759ff.). The authoritarian subject completes his social adaptation by drawing benefit from obedience and subordination. The Jew and every racially inferior subject became substitutes for a hated father, since in this fantasy they assumed those features of the father against which the subject had rebelled, such as being "practical, cold, domineering, and even a sexual rival" (ibid., p. 759).

In reconstructing the variables that could be traced back to the Fascist scale (F-scale for short), Adorno listed several psychic traits (ibid., p. 228ff.): conventionalism, in the form of strict adherence to conventional middle-class values; submissiveness to authority; authoritarian aggressiveness, which took the form of spying, punishing, condemning, and rejecting all those who violated conventional values; opposition to imaginative individuals and tender souls (and so to feminine traits); superstition; toughness and the exercise of power, with a constant concern with domination and submission; identification with power figures and an excessive importance attributed to ego attributes; an exaggerated assertion of strength and hardiness; destructiveness and cynicism; and an exaggerated preoccupation with sexual acts and projectivity, that is, the disposition to believe, on the basis of emotional and unconscious impulses, that dangerous events take place in the world.

Also stereotypical were those who scored low on the authoritarian scale (F-scale), and who developed the opposite tendency toward *rebellion* (ibid., p. 726ff.). The *protester*, too, seemed to respond to instinctual rather than rational determinants. The Oedipus complex profoundly influenced protesters, who, while being hostile to paternal authority, had at the same time internalized the image of a high-ranking father. The Superego was so strong in them that it was directed against the paternal model and against any external authority. These subjects were completely guided by conscience, but the reaction against prejudice seemed to have been imposed, in a sense, by the demands of a rigid Superego (ibid., p. 774ff.). The psychoanalytic explanation also informed Frank's (1949) legal realism. Our attitude toward the law was determined by the belief that law was a surrogate of the paternal figure: the adult resorted to law for fear of change and incontrollability (Frank 1949, p. 19). Therefore, the myth of legal certainty depended on our rediscovery or search for the paternal figure, who was in a position similar to that of a stern judge.

In the process of discriminating between "rational action" and "rational result," and in describing situations of "limited rationality," it is also essential to analyze actions and structures in public and private organizations. Here, too, we can hypothesize that individual attitudes regarding the sense of equity and justice lead to substantial distortions in decision-making processes, which we cannot analyze and solve according to the usual rational/irrational distinction, but rather according to a *conventionally* rational versus*differently* rational distinction, or, as Campanale (2007, p. 13) puts it, a distinction between the "conventionally" and the "humanly" rational.

The starting point can only be Weber's theory: organized rationality and the preponderance it plays in the modern world emerge in his political writings (Weber 1994). As is known, the primacy of administration over politics, and the progressive depletion of the latter's powers, is regarded by Weber as an inevitable fate of society. The action of the bureaucracy is driven by two fundamental principles: (1) "formalism" and (2) the commitment of officials to carry out their tasks in a "materialistically utilitarian" way for the welfare of their subjects (Weber 1978, p. 217ff.). The process of "de-emotionalization" of administration reaches its apex with the modern bureaucratic power structure, "making it so that the rationalization of the political sphere eventuates in the systematic exoneration of institutions from the emotional accidentality of the individual personalities that embody it" (Fitzi 2011, p. 45). However, the legal rationality of the bureaucracy is a formal condition that does not necessarily satisfy the material need for bureaucratic efficiency, and hence the collective interest.

Leaving aside the numerous controversial definitions of bureaucracy, such as "trained incapacity" or "occupational psychosis," we see that in sociological theorizing, organizational dysfunction is directly caused by deviation from legal rationality. The problem of bureaucracy is defined by Merton (1968, p. 260) as one of the "Solomon's Houses" of modern sociology. Paradigmatic in this respect is the analysis of the dysfunctional traits of bureaucratic personality, summarized by Merton in (1) ritualism, (2) esprit de corps, and (3) impersonality (ibid., pp. 254–255).

As far as ritualism is concerned, what happens in organizations is that adherence to legal rules, originally conceived as a means to an end, becomes an end in itself, this through a transposition process by which instrumental values become intrinsic values. Strict adherence to norms and regulations leads to an inability to adapt to situations that are not contemplated (ibid., pp. 177–178, 156–158). The transposition of feelings from goal to means is favored by the symbolic meaning of regulations. In this case, the attitude toward the norm seems to satisfy a need for stability and for conservation of the status quo, a need that is often nonconscious. This attitude can also be found in the problem of a public official's unaccountability. According to Merton, the bureaucrat tends to pigeonhole the individual problem of the citizen following a predetermined mental pattern that reduces complexity, and for this reason he is not able to understand its peculiarities, thus giving rise to tensions and dissatisfaction. That is different from the attitude toward the *esprit du corps* issue, where public officials, by leveraging their broad discretion in the enforcement of norms, take advantage of the interpretive freedom to make their status more politically salient.

In a study of the French bureaucracy of the 1950s and 1960s, Michel Crozier (2010) offers one of the most significant contributions on the distortions of rationality in the public sector. Crozier starts from the observation that in every organization of a certain size, decision-making shifts upward in the hierarchy for no functional reason. This shift toward the top solves the problem of the tension between executives and subordinates, because decision-making is entrusted to the top tier, but it creates a bottleneck, since the organization's leaders often do not have the time or the knowledge needed to make a good decision. This defensive dynamic soon turns into a "stable vicious circle": in order to put a remedy to a problem, the organization makes it worse, losing any self-correcting ability. With this instinctive mechanism of pacification and attenuation of conflict (declining a decision-making role to avoid a situation of tension) comes the paradoxical effect of a defensive horizontal solidarity. Peer pressure protects the official against the risk of total involvement in the organization's goals, and this in turn leads to corporative impersonality.

It is no accident that the question of organizational rationality came to maturity in the social sciences. This was owed to the work of scholars who took on the problem of human behavior from different perspectives. A case in point is Herbert Simon (1961), who in his most insightful works underlines how the practice of aggregating complex organizations serves the purpose of supplanting limited human rationality, and distinguishes between procedural and substantive rationality (Simon 2007, p. 35). While individuals cannot consider all available alternatives, administrative organizations are conducive to rationality: they support decision-making through predefined training structures, procedures, programs, and orders. Certain limits, however, cannot be overcome even through organizational structure, which can neither completely separate the means from the ends (considering that means are not neutral) nor anticipate all the consequences of a choice. Given its complex structure, the organization in any event requires compromises and adjustments in its decision-making. Simon's theory lacks a specific analysis devoted to law, despite the fact that legal epistemology could have provided an important angle from which to investigate the question. The evolution of the way law is represented is exemplified by the transition from perfect rationality to an imperfect rationality tempered by reasonableness, and plausible attempts have been made to find parallels in Luhmann's paradigm of complexity reduction (Campanale 2007, pp. 18–19).

Limited or bounded rationality is an inevitable constraint. The very methods used to remedy irrationality are subject to the same deviations. Collective decisions are themselves poorly rational, resembling an "organized anarchy" (Cohen et al. 1972, p. 1). The ambiguities that affect the decision-making process also affect the actors' preferences, the technology and procedures used, and the involvement of the actors in the decision-making process. As for these actors, the goals and expectations they have relative to the matter to be decided are unclear, because they are under-qualified, or because their experience is not pertinent to the decision, or because the goals are formed during the same interaction and change over its course. The organization, which should rationally guide the subject, becomes a set of choices and solutions in search of problems, feelings in search of situations in which to express themselves, and decision-makers in search of issues to be decided. Problems, solutions, and participants are discarded, then remain on hold, and then are unexpectedly and urgently recovered when an emergency comes up: they continually cycle in and out, as from a "garbage can" (Cohen et al. 1972, p. 5) into which they are thrown and from which they are retrieved, and the connection between problems and their solutions is often purely casual. Cognitive confusion reigns! Decision-making, whether or not mediated by organizations, has random features that separate it from rationality as commonly understood.

Theories of public choice (Niskanen 1971) assume that organizations are run by the opportunistic and maximizing bureaucrat, who follows either of two behavioral models. The budget-maximizing bureaucrat attempts to maximize the budget of his or her office to indirectly benefit from it by increasing its size and spending power. This does not result in direct economic profit, but it definitely leads to prestige and self-aggrandizement. The bureau-shaping bureaucrat does not look to acquire resources directly but rather aims to shape his office, emphasizing high professionalism and exploiting his position, close to political power and to the elites.

The final crisis of the Cartesian rational model has been reaffirmed more recently by challenging the concepts of subject and conscience (Crespi 1989, p. 15). "Action almost always encompasses […] elements whose meaning needs to be understood by reference to problems of identity, honor, or prestige, or to the symbolic force that certain objects or events take on in the dynamic of desire and of individual and collective imagination" (ibid., p. 127). Crespi questions whether broadening the concept of rationality does not engender contradictions that make the same concept outdated and useless. Von Hayek (2013, p. 10ff.) attributes the mistakes of constructivist rationalism to Cartesian dualism. What is certain is that there is no rationality separate from other abilities such as perception, thought, and language (Searle 2001, p. 22).

Aside from some vague acknowledgements, the social sciences and legal theory still do not seem to have drawn conclusions from this paradigm shift or from the back and forth between reason and cognition as fundamental schemes by which to interpret human behavior. Luckily, a decisive step toward a sociological perspective freed from the spurious rational-irrational dilemma has been taken with the analysis that social science has devoted to the emotions.

1.4 Emotions and Justice

The awareness that passions-affections-emotions are precisely those decisive variables that need to be studied, if we are to fully understand the social phenomenon, is widely anticipated in various passages of Hume's *Treatise of Human Nature*, which can be considered the foundational study of the modern cognitive sciences: "Reason is, and ought only to be the slave of the passions, and can never pretend to any other office than to serve and obey them" (Hume 1739, pp. 2, III, 3). Pareto, too, clearly understood the relevance of affective states in the study of social phenomena: "It is studying emotions that matters, in order to discover the laws of the development of society […]. Everywhere you find similar phenomena, everywhere, below the illogical and sometimes absurd derivations, you will find these feelings that are the great forces from which the form and the development of society derive" (Pareto 1980, pp. 580–581).

The concept of emotion I understand in a broad sense, including within it not only short and intense events but also longer-term moods such as sadness or happiness, and abiding feelings such as friendship and love (Pannarale 2008; Sorokin 1950). From a biological perspective, emotions are a step forward by comparison with "fight or flight" reactive mechanisms. In mammals, and especially in primates, emotions regulate attention, interest, motivation, and memory; they are older than language as a means of communicating with oneself and with others in society. Emotions have long been mistakenly confused with instinctive reactive mechanisms, and so they have been analyzed with improper methodologies. Stressful emotions induced on laboratory animals, for instance, were considered analogous to human ones, and it has been concluded that their simplest functions were also their only functions (Wentworth and Yardley 1994, p. 5).

Only recently has a systematic analysis of emotions in sociological theory been developed (Franks 2010; Turnaturi 1995; Wentworth and Ryan 1994), but their importance is now clearly acknowledged: "no action can occur in a society without emotional involvement" (Barbalet 2002, p. 2). Following the path of social psychology with considerable delay (Harré 1992), sociology has long resisted recognizing the "emotional man" as relevant in a true sociological study of emotions (Iagulli 2011, p. 42). Indeed, if relationships between people are also emotional, feelings and emotions should receive some emphasis not only in the sociology of everyday life, but also at the social level in its broadest sense (Iagulli 2011, p. 17). Collins outlines a real theory of emotions by highlighting their ritual value. The rituals of interaction, of encounter, of shared social situations, with the emotional component that characterizes them, ensure the ability to hold together. Social behavior is based on a motivational force that is given by the energy, trust, enthusiasm, and emotional warmth provided by individuals (Collins 1988, p. 362).

Once we were past the phase of intellectualizing of emotions, emotions became "social things" (McCarthy 1989) and "cultural objects" (McCarthy 1994) and took on an explicit meaning they previously lacked. As things, they become a reference point and an object of action, but they are also used to communicate something to oneself or to others, to take positions or to assume a role (McCarthy 1994, p. 275). This newfound line of inquiry came to the point of announcing the discovery of an "emotional culture" (Turnaturi 1998, p. 227) and defining dynamics such as "emotional deviance" (Thoits 1995, p. 124). Leaving aside evocative definitions such as the "post-emotional society," the reasons why emotions gained new interest are due to two orders of factors (Williams 2001, pp. 9–12): (1) the rediscovery of the Dionysian dimension in contemporary social life, where a collective and communitarian empathy manifests itself in the decline of individualism and in the rise of neo-tribalism, and (2) a renewed interest in the role of biology in explaining the social.

The sociological study of emotions, however, does not have solid foundations yet, and a sociologist who is not of strict ethnomethodological adherence does not feel entirely comfortable with the topic. A conditioned response persists that leads sociology to label as behaviorist psychologism any discourse that is not at least in some way *prima facie* related to the macro dimension. The classic sociological approach dealt with emotions indirectly and unsystematically. It is correct to say that only the phenomenology of emotions has now been satisfactorily completed, and we do not yet have a proper sociological theory of emotions. But if it is true that emotions often deviate from a rational behavior, they function as "an engine and medium of many social exchanges, while making it possible to deal with our cognitive limits and decision-making impasse in many choice processes" (Mutti 1994, p. 74).

The social value of emotions, however, is neither symbolic nor subjective, because emotions have an intrinsic social value that does not depend on the social value of our description of them. As much as emotions have a biochemical basis, they are not purely naturalistic. They are culturally specific, and so cannot be reduced to mere social conditioning (Iagulli 2011, p. 7). In pointing this out, I intend to reject a dramaturgical or interactionist interpretation of emotional sociality, and I fully agree with Williams when he states that while emotions are a culturally mediated element, they are also a structured and structuring social force in their own right (Williams 2001, p. 49). The emotional actor is a complex actor, self-reflective and not at all irrational or tied up only in the immediate impetus (Turnaturi 1998, p. 239). Emotions are socially effective and constitute an index of social ties. They operate at the core of normal routine in everyday social processes, including social order structures and harmonious social change.

Among the first scholars to thematize the problem of self-imposed and external constraints that modify the expression of emotions is Norbert Elias (2000). Elias explains how biological and social factors intersect in the experience of feeling and communicating emotions, bringing the biological into the social, while recognizing that there are no entirely instinctual emotions that do not have at least some learned element. In evolutionary terms, we humans could be an absolute novelty, in that our learned behaviors outweigh our instinctual ones. The ability of humans to learn has almost become a necessity. Even if ethology may question the assumption of human uniqueness and also highlight the importance of learned behaviors in nonhuman beings, this is still a remarkable break with sociological constructivism, which places emotions in linguistic habits and contextualized language games, and for which emotions are almost entirely unnatural (Williams 2001, p. 54).

The continuing and still unfinished construction of a sociological theory of emotions is based on the ethnomethodology of daily interaction, but it will have to distance itself from that kind of investigation. In this respect, one fundamental perspective is that of *role-taking emotions*, considering the function they play in the construction of social control (Shott 1979, p. 1317). Many human emotions are elicited by real, anticipated, or even imagined outcomes of social relationships. Role-taking emotions cannot be felt if not by taking the perspective of another person.

These emotions, which are strongly relevant to the social relationship, have been divided into two types: (1) reflective role-taking emotions, which are directed at oneself and include guilt, shame, embarrassment, pride, and vanity, and (2) empathic role-taking emotions, which are experienced and invoked by putting oneself mentally into someone else's shoes, in order to feel what they would experience in that same position (Shott 1979, p. 1324). Role-taking emotions facilitate social control by forestalling deviant behaviors and favoring self-control, and they would be the reason why, individually, we mostly comply with norms even without significant punishments or rewards (Shott 1979, p. 1329). Another peculiar example of a role-taking emotion is the *lek*, which is specific to Balinese society, but which can be counted among the "affective regulation" tools. It is an emotion similar to stage anxiety: an all-pervasive sense of nervousness, usually mild but sometimes paralyzing, that precedes a social relationship or the expectation of it (Shott 1979, p. 1326).

As much as Weber's analysis may not be useful to us in the attempt to reconstruct the subjective cognition of action, it is certainly true that the emotions are understood by Weber as an integral part of social reality, even if he has been faulted for having a "rationalist prejudice" (Fitzi 2011, p. 37). Affective action is conceived by Weber as a sudden emotional reaction to the unexpected (Fitzi 2011, p. 38). And, as noted, the sociological meaning of affective states remains on the margins: simple mimicking or mass action is normally excluded from interpretive sociology. The outbreak of the passions becomes sociologically relevant because it affects action or social structure. The relevance of habit as distinct from calculation is decreasing in a modern society. But traditional habits, as well as violent passion, are still "the fuel that serves to refashion social relationships into something new" (Fitzi 2011, p. 49).

Although Weber appears to take the view that rationalization and bureaucratization tone down charisma (Iagulli 2011, p. 31), he does not predict an extinction of the emotional side in favor of a complete rationalization. He even concedes, in passing, the permanence of one emotional (affective or habitual) side which acts as a catalyst through which to periodically escape from the "iron cage," and which promotes evolutionary leaps in social becoming. The sociology of sentiments runs through Weber's entire political, economic, and religious theory. This can especially be appreciated in his political analysis of the personal and extraordinary charisma of the chosen leader or ruler, the great demagogue, and the party leader (Weber 1978, pp. 215–216, 241ff., 1111ff.).

There is notable heuristic validity that Weber (1978, p. 40ff.) sees in the social process of being in a commonality (*Vergemeinschaftung*): this process gives us a sense of belonging, and in particular it can be identified with the *Emotionalen Vergemeinschaftung* (emotive or emotional commonality), which acts as a group force in traumatic and irrational collective events. Although charisma and emotional communion are of limited duration, they can affect social relationships, but they are destined to fade and are not meant to last. Quite different is the status of agapic love (Tammelo 1957, p. 383), which also figures into this separation from the rationalist tradition that takes place with the sociology of emotions (Bettin Lattes 2011, p. 6). The relevance of love in modern sociology is primarily attributable to Georg Simmel (1984), according to whom love is one of the great formants of the existent. The reconstruction of the feeling of love takes place in the context of a renewed focus on emotional themes that appear to relate to the emotional, such as the relevance of fashion or flirtation, and that we would now trace back to the sociology of everyday life. As in Elias's work, so in Simmel's there is the perception that it is in the metropolis—the most sophisticated and evolved form of contemporary society—that a substantial change in emotional life happens.

The feeling of love is the basis for building sociality, where it represents a higher level of evolution: "The level of human development to which love belongs is too advanced for us to be able to situate it in the same temporal and genetic stratum as eating, breathing, or the sexual drive" (Simmel 1984, p. 161). In his work on love, Simmel reconstructs the forms of love as universal love or Christian love. Christian saints and good neighbors are the focus of another seminal research that calls for intensified study of the "energy of love" (Sorokin 1950, p. 231), but there is little else in sociology.

Not incidentally, Boltanski (1990) reframes the character of *agàpe* from a perspective that criticizes the prevalence of structure over the subject. Love as a source of social relationships means renouncing every measure by which to count how much is given and how much is received, on the assumption that everyone is unique and irreplaceable. Boltanski conceptualizes love as practical action rather than resolving it into feelings or moods. His "agapic-pragmatic" vision stands diametrically opposed to the macro- and hyper-conceptual determinism of his mentor Bourdieu.

Agapic love is completely selfless and cannot be the object of a discourse, theory, or project, whether personal or social. According to a different reconstruction, "*agàpe* is defined starting out from oneself and for oneself without interest, without expecting anything in return" (Iorio 2011, p. 108). Constructively, it can also be defined as a "gift stripped of any cause and free of any economic logic" (Iorio 2011, p. 109). It is in the experience of love that the subject can give meaning to a relationship with juxtaposed subjects, and it is therefore in love that the idea of an ethical community is internalized (Iorio 2011, p. 110).

In a recent work, Paolo Iagulli (2011) offers a valuable historical reconstruction of the theories that contribute to the still incomplete formation of a sociology of emotions. In his investigation, he first discards the constructionist perspective—which essentially denies that there is any biological aspect to emotion, considering the emotional part to be a socio-cultural construction—and thus leaves us with essentially two acceptable positions: (1) sociological models with a psycho-organic orientation analyze which social factors induce and stimulate emotions and feelings, and (2) psycho-interactionst sociological models study secondary actions performed during a primary emotional experience (Iagulli 2011, p. 51ff.). The positivist perspective contends that emotions are determined by power relations and status. In some versions, this perspective also incorporates the additional motivation factor given by a cognitive assessment of the situation, of expectations, and of sanctions. Human emotionality is the result of social relationships, even if only predicted or imagined.

To this end, if sociology is to understand emotional phenomena, it needs to work with physiology and psychology. Positivism thus moves closer to behaviorism and, while it does not disregard the influence of socio-cultural components in emotions, now reduced to activation factors, it defines emotions as physiologically determined and measurable, hence predictable, objects. The psycho-interactionist approach instead tries to show how people conform in outward appearance but never in their inward experience (Iagulli 2011, p. 53). It shows us the result of suppression but not the process or the techniques through which this is reached. Among the cognitively most relevant emotions is that of understanding (or sympathy), which consists in the emotional-cognitive process of supporting and comforting the other, but without identifying with his or her position (for that would be empathy: see Sect. 4.3). The exchange dynamic is rooted in strong cultural elements: those who *offer* understanding do so because they feel obliged by a set of cultural norms, while those who *receive* understanding feel obliged to reciprocate with manifest gratitude and relief (Iagulli 2011, pp. 79–80).

Whether someone is particularly worthy of understanding or is expected to show understanding depends on a set of different cultural parameters: people with special burdens, those who have a special responsibility by virtue of their exceptional qualities, subjects who are weak owing to a vulnerability or to the potential that their existence still has to express (children, for example). The cognitive perspective becomes relevant in the study of emotions, especially if their role is fully understood: by giving weight to those sensory perceptions that stand out as more relevant among the innumerable perceptions we constantly process, emotions signal those perceptions that require urgent attention. It is functionally useful to distinguish affection and cognition, as these should be considered symbiotic partners rather than antagonists (Greenberg and Cohen 1982, p. 83).

No less interesting is the classification offered of negative emotions such as resentment and shame. Resentment emerges when individuals perceive that some have received material gains in ways that are deemed undeserved according to norms of justice or cultural expectation. Resentment is more common in an inter-class relationship, but it is not excluded among individuals in the same class, since it can originate from any perception that certain individuals have used their power to deny others their proper status. It may be conscious or unconscious, leading to criminal acts of revenge or to collective mobilization (Iagulli 2011, p. 83). Shame favors the consolidation of social ties when it is recognized and used to remedy a lack of social approval. It turns instead into negative and potentially destructive shame/loathing cycles when repressed. Also famous is the example of the *piacular rites* of mourning: a conformist and collective expression of emotions for ritual purposes (Durkheim 1912, p. 289).

The emotional perspective is also employed in relational theories, which propose a synthesis between sociologies of actions and sociologies of systems, proposing to show that neither is complete. Simmel (1982, p. 17) defines an object as unitary as long as reciprocal dynamic relationships exist among its parts. The fundamental philosophical dualism that differentiates the individual as subject from the individual as object is rejected by a *critical relational realism* (Donati 1983, pp. 11–12). Relational sociology is linked to philosophical anthropology, and it aims to go beyond empirical social research (ibid., p. 214). The criticisms that relational sociology devotes to the mechanisms of "social reproduction"—perpetuated in critical paradigms, ethnomethodology, interactionism, system theories, and Marxian neo-positivism—point out all the theoretical errors and *sociologistic* contradictions involved in theorizing these mechanisms. The first target is Marxian constructivism, which with Althusser and Bourdieu has come to a hyper-functionalist short-circuit by denying any agency to the human subject: in this scheme, social groups act simply as performers of roles defined in the class structure, in a fully deterministic way and hypostatizing the social system along with its relationships.

To better illustrate the fundamentals of relational sociology, Emirbayer takes a substantialist perspective. This perspective proceeds from the idea that the fundamental units of each inquiry are substantial: they are things, beings, or essences (Emirbayer 1997, p. 282). Society is best described as a complex of relationships understood as typical real connections in a non-reductionist and super-functional sense. The social relationship cannot be restricted to the functions it performs. Donati argues that sociology is also an applied science that cannot but legitimize the existent, but in its humanistic dimension—one that is historical but not historicistic, for it does not assimilate social relationships to the historical forms they take—it aims to uncover a fully accomplished humanity (Donati 1983, p. 198). Relational sociology places itself between irrationalist and hyper-rationalist sociologies, both of which are the result of the "cultural colonization" of mature capitalism (ibid., p. 215).

Relational sociology could also be a valid alternative to rational-choice theories, which take individual human action as the elementary unit of social life, using it to explain social change and the emergence of social institutions (Emirbayer 1997, p. 284). The success of rational choice originated from the rigor and sophistication of the game models used to empirically validate its laws. However, the postulate of rationality implies that any other behavior should be interpreted as a derivation from that rational scheme, and that only pure models of rationality can be used in science. And yet scientific analysis cannot be validly applied only to the "puppets" or the "homunculus" the scientist has created or posited for experimental purposes. Using a consistent principle of adequacy as our guide, we can easily find some types of "instrumental non-rational actions" in the social world—examples being expressive actions, traditional actions, or value-rational actions—which require proper conceptualization, and which do not correspond to a pure model of rationality (Donati 1983, p. 215). The explanation of a social phenomenon may consist in showing that it is the unintended effect of rational actions (Boudon 2009, p. V) or that small institutional changes may have consequential effects that cannot be qualified as either negative or as positive: they are simply "opposite" (ibid., p. VIII) or, in other words, as "perverse" (ibid., p. 7).

The field of normativity is closely linked to that of the emotions (Mutti 1994, p. 77), and the field of emotionality and the passions begins to also express itself in the doctrinal study of law (Bandes 1999; Massaro 1999), where emotionality had hitherto been regarded as an exceptional state, to be taken into account only as an aggravating, mitigating, or justifying factor in the given circumstances in determining the penalty for a violent offense, or where it was taken to be one of the causes of mistake by a contracting party. This relationship is one whose acknowledgment puts the modern positivist conception of law in a tight spot. The problem of the emotional charge inherent in the acts taken into account by law, and how to deal emotionally with the way the law solves this problem, in turn gives rise to a range of other problems whose solution is elusive (Nussbaum 1999, 2004).

To what extent can the law's assessment of a fact take into account the degree to which emotion has played a part in the act subject to punishment? Can law use emotion, and in what way? When and up to what point should the emotions of legal professionals and officials (judges, prosecutors, clerks, lawyers, police officers) be considered relevant, and what measures can be taken to ensure that they carry out their tasks in the correct emotional state, whatever it may be? Posner (1999) seems to conclude that the law cannot be directed—whether in the sense of exalting the emotions or in that of ignoring them—but the compromise solution does not solve the practical problems of legal interpretation, as when, in a courtroom, we are faced with the problem of the disgust at a heinous act or at the person who committed it (Nussbaum 1999, 2004).

Nussbaum argues that anger and indignation are acceptable feelings in legal reasoning (Nussbaum 1999, 2004). It has insightfully been noted that the problem of disgust emerges *ad adiuvandum* in socially conservative legal arguments that defend the preservation of traditional moral values and, symmetrically, in liberal critiques of the same arguments (Kahan 1999, p. 63), but it would be useful and appropriate to also distinguish disgust from contiguous emotions (Posner 1999, 2004) which are distinct from distaste or the sense of danger, but which sometimes result in similar behavioral outcomes. For example, Nussbaum warns against the feeling of disgust, which negatively affects the legal evaluation of a fact. Posner (1999, 2004), on the other hand, asserts its importance: disgust enables the judge to take into account the needs of parties who often cannot participate in the trial (in cases that elicit horror and disgust, these are often the dead victims), for it gives greater emphasis to their suffering and thus counteracts the availability bias, that is, the heuristic that forces us to overvalue and thus favor elements that are "cognitively" at hand (in these cases, the defense of the accused, and the need to understand his or her reasons, or the state of affairs that led to the commission of the act). It should be acknowledged that when a prosecutor carries out his or her role empathically, and there is an opportunity for the victim to take part in the trial, the risk of the judge falling victim to the availability bias should be negligible. Disgust cannot simply be discarded but must be actively managed before it can be used effectively in a judicial decision: the impossibility for society to vent its feelings institutionally ultimately leads to populism and witch-hunting.

Revenge is considered a more complex impulse: it finds a rational explanation within a moral and retributivist perspective, and is often mingled with the most primary emotions of resentment and hatred (Girard 1986). It was a codified practice in ancient law, but it has also made its way into modern forms, and only recently was it marginalized and expunged from the legal system (French 2001; Lorini and Masia 2015; Pigliaru 1959; Terradas Saborit 2008). The human desire for punishment and retribution does not demand any philosophical rationalization (Jacoby 1983, p. 9). Leaving aside the position that vengeance would be the one and only purpose of criminal punishment, even when masked by principles of crime prevention and re-education, it has been argued that the law cannot simply ignore the desire for revenge but must strive to trace it back to a retributivist rationality (Solomon 1999, p. 127). Likewise, the *lex talionis* represented an effort to bring the instinct of indiscriminate retaliation back within the bounds of the rationality of retribution, and it seems appropriate to draw a distinction among revenge, retribution, and retaliation (Solomon 1999, p. 131ff.).

Also noteworthy is Posner's (1999, p. 326) criticism of the modern moral rigor that emerges in this debate between forgiveness and revenge. Jacoby (1983, p. 13) argues that the taboo that civilized societies today attach to vengeance carries a risky potential for social regression, for it acts contrary to the self-control it is supposed to encourage. In some cases, however, the courts themselves still use revenge selectively and improperly as a mitigating element, or at least this is what a study of "sexual vengeance" (Jacoby 1983, p. 183ff.) has found in cases involving crimes of passion, especially infidelity, be it actual or potential. However, as evidenced by the public outrage at *femicide*, the tide is turning in what concerns the social perception that revenge is understandable when motivated by sentimental reasons: it is giving way to a common perception that presumably now results in aggravated circumstances, or in any case in greater social disdain.

The feeling and perception of law and rights are measured in the first place, and perhaps exclusively, by what we perceive to be fair. In the social sciences, the idea and sentiment of justice have been made to coincide with the content of the law, this on the assumption that these were reserved for ethical and philosophical analysis, entrusted with identifying its substantive contents. With the development of psychology and social psychology the disconnection problem is accentuated. The idea that feelings of justice drive behaviors and stabilize social relationships is ignored in Homans's theory of social exchange, and is pursued in social psychology (Arts and van der Veen 1992, p. 155).

The study of individual motives and subjective experience as they relate to justice adds a further descriptive and experimental approach and contributes to the fragmentation of research on this issue. The occasional and increasingly rare contacts between disciplines are seldom fruitful, not only because of the terminological distance but, more worryingly, because of the divergent methodological preconditions, if not for the underlying conflicting visions of the world, or even because of "mutual ignorance and arrogance" (Bell and Schokkaert 1992, p. 252). The discouraging outcome is the incommensurability of theories about justice (Cullen 1992, p. 60).

Theoretical sociology focuses on the observation of the effects of justice or injustice on society and social movements. Justice therefore remains an implicit object of study in sociology, with the *sociology of justice*, of Kelsenian origin, representing its specific branch in the sociology of law (Di Lucia and Passerini Glazel 2015). The idea of justice is connatural to social life itself, representing both "an artifice or contrivance, which arises from the circumstances and necessity of mankind" (Hume 1739, pp. 3, II, 1) and a sentiment whose violation is the object of resentment and punishment (Smith 2002, p. 93), and which "comprehends all the social virtues" (ibid., p. 319). Huxley (1989, p. 115) points out the historical evolution of this concept, gradually sublimated from punishment or reward for actions to punishment or reward depending on merit or intent. Weber, in all his works, limits his analysis of justice to its historical evolution, to the assumptions and the functions of law in society, which is to say to justice as the administration of justice, as a further specification of the expansive principle following which human societies are increasingly tied to the organizational form of bureaucracy.

According to Blau (1964a, b), the origin of the idea of justice is to be reconnected with social justice, and justice is implicated in the relation between social needs for the fulfillment of specific tasks, on one side, and the investment needed by people and groups for the success of these social functions, on the other. The search for the concept of justice, "whatever its success, represents a social-psychological phenomenon of great importance, especially in a modern democratic society. It sums up the attitude of constant criticism of existing law, and of unwillingness to accept the law for the time being as more than means to other ends" (Stone 1966, p. 784). One can see here a first explanation of the attention that socio-legal scholars pay to procedural justice, while general sociology instead seems to be more interested in the substantive aspect of distributive justice, and so in social stratification and inequality (Arts and van der Veen 1992, pp. 144–145).

Even in Luhmann's hyper-functionalist view, which separates justice as a process from justice as an ideal, it is suggested that the indeterminacy of goals in modern complex societies that rely on economic laws can only provide a critical and evolutionary vision of justice, rather than a perfectionist and absolute one (Bell 1992, p. 124). The aim of theory is to find "socially appropriate legal concepts," but this does not mean that they are sociological concepts or that they adequately reflect society (Luhmann 1978, p. 107).

It is true, however, that the sociology of law is concerned with justice, for it reflects on questions relating to the legitimization and foundation of law (be it formal-procedural or material-substantive) and to the complexity and adequacy of the legal system, or ultimately of the legal culture and its impact on social action. Even socio-legal research explicitly concerned with the social determinants of the concept of justice (Eckhoff 1974; Ferrari 1995), though interested in analyzing the socio-demographic variables that could influence the choice between alternative criteria, takes the traditional Aristotelian distinction between commutative and distributive justice as its starting point. The distinction can be found in Book V of the *Nicomachean Ethics*, where different kinds of justice also receive other designations, such as *corrective* or *retributive justice*, depending on the translation. Strahlendorf (1992, p. 135ff.), for example, distinguishes between distributive, commutative, remunerative, and procedural justice.

Although Aristotelian distinctions anticipate the more recent distinctions between formal and concrete justice (Perelman 1959, p. 57ff.), or between procedural and distributive/social justice (Cullen 1992, p. 16), sociology is closely tied to this ethical-philosophical approach, for in its analysis the variable of the ethical or systemic function of justice is deemed indispensable. Even Luhmann's previously mentioned perspective—in its transcending ethical values so as to focus exclusively on the values of the system's balance and survival through "stabilizing expectations of behaviors" (Luhmann 1978, p. 59)—does not differ methodologically from the most liberalist and subjectivist views. In summary, the description of justice is a moral problem for philosophers, a problem of efficient implementation of social choices for economists, a problem of compliance with rules for jurists, a problem of individual reactions in interpersonal relationships for social psychologists, and a combination of all these problems for sociologists (Bell and Schokkaert 1992, p. 237).

In this last assertion I see an opportunity for sociology to act as a collector of epistemological visions of justice made functional to a predominant goal. The ethical theories may be able to guide and relativize the results of empirical research, while the empirical approaches, simply by observing regularities, can in turn inspire ethical constructions, recognizing the difficulty of finding universal truths in all aspects of social life (Markovsky 1991). I, too, believe that the scientific debate on what the most equitable economic system is, for example, cannot ignore the complex cognitive factors affecting the perception of justice or the social factors affecting those cognitive factors (ibid.). If we are to craft effective social policies, we need to be able to understand the cognitive and social factors that shape the underlying subjective evaluations.

As Scherer (1992, p. 9) suggests, the achievable medium-term result, considering the difficulty of the concept and the relativity of values, consists in "defining a procedure to adopt in the discussion of justice questions and relinquishing the dream of being able to agree on universal principles of justice." In doing so, sociology must also pursue other avenues, and not shrink from investigating those subjective visions of justice that explicitly exclude ethical and prescriptive elements.

1.5 Toward an Empirical and Physicalist "Cognitive Sociology"

In the rest of the discussion, I will outline the methodological foundations on which to set a *cognitive sociology of law* that can integrate the contributions of different social and natural disciplines, without merely stacking them up by additive accretion, so as to finally arrive at a hypothesis about individual social attitudes to the law. This will make for an evident impact on the sociological study. Sociology has already analyzed cognitive expectations about the law, for example in its description of law as a selective scheme that, through learning and decision-making strategies, reduces the complexity of individual expectations (Luhmann 1978). Now we need to describe what these strategies are and how they operate at the individual level.

Although it has been shown that the distinction between the macro and the micro has in many cases been used to set up false problems that might be solved through a synthesis, the approach proposed here is surely closer to a microsociology inspired by an empirical and "physicalist" methodological individualism (Meyering 2000). A microsociological and individualistic option so described is obviously subject to criticism. The obsession of sociology has always been to find the ultimate social factor of change, the "bearer of historicity," which by turns has been located in class struggle, technological advancement, the mass media, molecular biology, the globalization of companies, and the intellectual class, to name but a few (Boudon 2009, pp. 17–19). Paradoxically, it is perhaps the unyielding search for universals that has led us down a tortuous path. Functionalism, too, has led to untestable universals, which only give us *pseudo-explanations* (Collins 1975, p. 6) or post factum rationalizations (Merton 1968, p. 147).

The search for universals, however, need not be unsuccessful when based on *testable* elements. Suffice it to mention, for example, recent hypotheses that look at climate change as a macroeconomic, social, and political factor by analyzing recent history on the basis of an extended body of climate data collected on the field, suggesting that a series of sudden variations have had a profound influence on secular political events (Behringer 2010). Or consider the idea that a society may be more individualistic or more collectivistic depending on the kind of farming practiced in it (Talhelm et al. 2014). No analysis of legal cognition can offer an all-encompassing explanation, leaving countless social behaviors and attitudes in other areas of social life unexplained.

However, all-encompassing explanations have decoupled the social sciences from the social world, creating between them that disconnect which has been frequently blamed on sociology (Schütz 1967). Without mentioning the *pure* macro level—which in its most extreme forms resolves itself into conceptual alliterations *à la* Bourdieu (Bourdieu 1981)—I share the view that many systemic methodological frameworks, and not just those of legal sociology, are too abstract to yield verifiable hypotheses in this as in other areas of human activity. Some have wondered whether these conceptualizations fully satisfy the canons of science (Tamanaha 1997, p. 19). Since I am taking a stance against holistic explanations in the social sciences (Popper 1971), I find it hardly acceptable, as a natural consequence, to consider the intentions and actions of social agents always as effects and never as causes (Boudon 2009, p. 15).

I am not persuaded by the attempts made to define sociological inquiry in an anti-Popperian sense, especially when they originate from a range of perspectives whose only common denominator lies in their criticism of "experimentalism," paradoxically accused of promoting an unconditional primacy of theory (Passeron 2006, pp. 539–540). Here Passeron summarizes the anti-experimentalist thesis by arguing that there cannot exist a "protocol language" unified by an empirical description of the "historical world," and that in sociology, the empirical proof of a theoretical proposition cannot take the logical form of refutation (falsification) in the Popperian sense.

I also share the view that a macro-approach is indispensable, so long as it is supported by an awareness that the theoretical framework it produces—based on macro-concepts such as the system or society—is necessarily incomplete and needs to be integrated to respond to the present-day demand for predictability (Collins 1981, p. 94). In the analysis of individual attitudes to the law, the macrosociological paradigms are wanting in several respects: in their disregarding the contingent nature of social interaction, in their forcing social reality into predetermined boxes, and in their ignoring the limitations of the cognitive abilities, which sociologists themselves have in common with the social subjects they study. As Collins (1981, p. 85) points out, the accusation of "subjectivism" backfires against those who use it against individualist paradigms.

In view of the task I have in mind, however, the theoretical paradigms that symbolic interactionism and sociolinguistics have led to do not seem entirely adequate. As we have seen, individual attitudes toward social phenomena have been widely explored, when the "cognitive turn" of the 1950s gave rise to ethnomethodology, social phenomenology, and a variety of other approaches that focused on everyday social interaction (Knorr-Cetina 1981, pp. 1–2). Microsociology became increasingly radical even from a philosophical point of view, especially with the ethnomethodological approach, and it has moved beyond the pragmatism of a stipulative definition of the world, so as to build a true empirical phenomenology (Cicourel 1974). However, the linguistic-symbolic factor has come to be so preeminent as to crowd out and efface the properly cognitive aspects that preside over the interpretation and representation of values and obligations. Despite the fact that sociolinguistic theories have explicitly attempted to found a "cognitive sociology" (Cicourel 1974), the symbolization of the social act and its extreme relativization have led to poor results when it comes to predictive capacity. As a matter of fact, a paradigm that assumes that everything is extemporaneous interpretation barely allows any kind of prediction. It is unlikely that there is much to be gained, in tackling this problem, by taking up Collins's (1981, p. 95) suggestion that the term *reduction* be replaced by *micro-translation*. Cicourel's cognitive sociology has thus become part of the idealistic tradition, because it has been more concerned with thought than with action. It has been concerned with "putting a situational interpretive model under the old normative order rather than seeing human animals struggling over physical and symbolic territory" (Collins 1975, p. 111).

The microsociological option underlying this analysis needs to be tempered with an awareness that the most appropriate level of analysis in the social sciences is the one indicated by Merton, namely, the level of midrange theories, the level that is "intermediate to general theories of social systems which are too remote from particular classes of social behavior, organization and change to account for what is observed and to those detailed orderly descriptions of particulars that are not generalized at all" (Merton 1968, p. 38). Merton's analysis still deserves to be highlighted, because it seems that his lesson has yet to be understood.

The epistemologically immature state of sociology is borne out by the fact that specific sociological hypotheses derived from a central conceptual scheme have been few and fragmentary, and by the scarcity of confirmed theories, especially if we look at the proliferation of concepts, points of view, and approaches that have been produced: "To concentrate entirely on a master conceptual scheme for deriving all subsidiary theories is to risk producing twentieth-century sociological equivalents of the large philosophical systems of the past, with all their varied suggestiveness, their architectonic splendor, and their scientific sterility" (Merton 1968, p. 51).

I share this individualistic approach to the functional paradigm proposed by Merton, which embraces concepts and processes codified in the functional interpretations of sociology, social psychology, and social anthropology, all converging into a common theoretical framework. Sociologists need help from biologists, psychologists, historians, and philosophers (Runciman 1989). Even today it is worth Elias's invitation to "regain within a limited area the lost perception of [...] the peculiar transformation of human behavior," using "different branches of scholarship, which are often divided by artificial barriers today," like "psychology, philology, ethnology and anthropology no less than sociology or the different special branches of historical research" (Elias 2000, p. xiv). The same preoccupation was voiced by Gallino (1987) against the unwillingness of Italian sociology to work on a theory of the actor. In fact, I share the concern that conventional divisions in superficially distinct spheres of human behavior obscure the similarities between social and psychological processes (Boudon 1969, p. 449).

It may be useful to consider the parallel with the physical sciences, where incompatible paradigms, such as general relativity and gravitational theory, may coexist while waiting for the "theory of everything" that can explain both large-scale and small-scale physical events within a unified conceptual framework. Morin (1977, p. 15) stresses that while it is not yet necessary to look for an overarching or unifying theory, human science postulates natural science, which in a circular relationship postulates in turn human science, and he calls on sociology to go back to its original quest for a "social physics" (ibid., p. 284). What I am arguing here is that any science risks losing its scientific utility if it neglects to consider other disciplines. The warning is not new, considering that in the 1990s an old-school sociologist like Immanuel Wallerstein—directing the work of the Gubelnkian Commission, formed by humanists, social scientists, and hard scientists—criticized any disciplinary monopoly on the knowledge of social phenomena by invoking broader collaboration on specific subject areas such as population, health, and language (Wallerstein et al. 1996, p. 98).

Several investigations have been devoted to the idea of a sociological "theory of everything" that connects the sociological and psychological dimension: Simmel (1982, p. 13), for example, rejected the idea, while Collins (1987, p. 204) ventured the hypothesis that in looking for such a theory we should have to start from the concept of property. Meanwhile, we cannot content ourselves with erecting barriers to defend disciplinary purity. If this sociological turn is desirable, we must, as Tamanaha (1997, p. 90) suggests, look for a synthesis between interpretivism and behaviorism. Without a unifying "general frame of reference" comparable to the neo-Darwinian framework in biology, sociology cannot deliver on its promises (Lenski 2005, p. 222). The importance of finding an objective foundation common to the various paradigms was masterfully affirmed by Merton, when he argued that if sociology should not imitate chemistry, physics, or biology, neither should it only fashion itself after philosophical discourse and literature, renouncing the use of objective concepts in such a way that "the hard skeleton of fact, inference and theoretical conclusion becomes overlaid with the soft flesh of stylistic ornamentation" (Merton 1968, p. 70). In not too dissimilar a way, the same concept is echoed in the words of a geneticist who points out the tendency in sociology to prefer the construction of a *visually rich* picture to a *terminologically precise* one (Boncinelli 2009, p. 115).

What I am proposing is a cognitivist framework in which to combine the most fruitful insights of microsociology and anthropology with social psychology, evolutionary biology, and cognitive philosophy, in such a way as to arrive at an integrated description of law-influenced behavior and cognition and to formulate or clarify some hypotheses about legal attitudes. I do not intend to forge radically new definitions or concepts, and even in this I agree that the "road to effective general schemes in sociology will only become clogged if, as in the early days of sociology, each charismatic sociologist tries to develop his own general system of theory. The persistence of this practice can only make for the balkanization of sociology, with each principality governed by its own theoretical system" (Merton 1968, p. 51). The balkanization of sociology—and the sociologist's Columbus syndrome, that is, the bad habit of endlessly devising new labels and concepts where existing ones suffice (Elwell 2013, p. 10)—is now a sad reality, and I hope the completion of a "cognitive turn" could reverse the trend.

An empirical and blandly physicalist cognitive sociology may be useful in overcoming most of the objections of reductionism, behaviorism, and scientism, such as those that Dal Lago (1994, p. 62) has formulated in relation to individualistic methodologies. Society is real, as is cognition understood as a set of cognitive, analytical, and decision-making activities. However, the connections between cognition and society can only be analyzed on a multidisciplinary approach, and with at least some degree of unification between the social and the natural sciences. I do not intend to embrace a *strong* physicalism that conceives the social phenomenon as ultimately reducible to physical laws (while also going through the intermediate stages of psychology, biology, and chemistry) by way of bridge-laws that would strictly link disciplines from a given general level to a more specific one (Meyering

2000; Nagel 1974). This critique of strong physicalism suggests a broader and looser relationship between the sciences in general and physics in particular: this looser approach is referred to as *token physicalism* (Meyering 2000, p. 187) or, better yet, *compositional physicism* (Meyering 2000, p. 196). In this relationship, an identity of events does not necessarily imply a correspondence of explanatory propositions.

In essence, multidisciplinarity and convergence toward unified principles of explanation should not necessarily be pursued under absolute canons (something that cannot be achieved even within each individual science) but should rather be based on correspondences, assonances, and explanatory analogies. The cognitive sociology I propose here integrates the micro- and macro-perspectives by also bringing into the social sciences those human behavioral disciplines that help to clarify legal attitudes: this includes the philosophical and the cognitive vision in their phenomenological and nonsymbolic versions.

References

Addario, Nicolò. 1994a. Da Parsons oltre Parsons. Una soluzione epistemologia del nesso micro-macro. In *Il rapporto micro-macro nelle teorie sociologiche contemporanee*, ed. Nicolò Addario. Milano: FrancoAngeli.

———. 1994b. Il dibattito contemporaneo sul nesso micro-macro in sociologia. In *Il rapporto micro-macro nelle teorie sociologiche contemporanee*, ed. Nicolò Addario. Milano: FrancoAngeli.

Adorno, Theodor Ludwig, Else Frenkel-Brunswik, Daniel Levinson, and Sanford Nevitt. 1950. *The Authoritarian Personality*. New York: Harper & Brothers.

Alexander, Jeffrey C. 1987. Action and Its Environments. In *The Micro-Macro Link*, ed. Jeffrey C. Alexander, Bernhard Giesen, Richard Munch, and Neil J. Smelser. Berkeley, Los Angeles, London: University of California Press.

Alexander, Jeffrey C., and Bernhard Giesen. 1987. Introduction. In *The Micro-Macro Link*, ed. Jeffrey C. Alexander, Bernhard Giesen, Richard Munch, and Neil J. Smelser. Berkeley, Los Angeles, London: University of California Press.

Alexander, Jeffrey C., Bernhard Giesen, Richard Munch, and Neil J. Smelser, eds. 1987. *The Micro-Macro Link*. Berkeley, Los Angeles, London: University of California Press.

Arts, Wil, and Romke van der Veen. 1992. Sociological Approaches to Distributive and Procedural Justice. In *Justice. Interdisciplinary Perspectives*, ed. Klaus R. Scherer. Cambridge, UK: Cambridge University Press.

Baggio, Guido. 2015. *La mente bio-sociale. filosofia e psicologia in G.H. Mead*. Pisa: Edizioni ETS.

Bandes, Susan, ed. 1999. *The Passions of Law*. New York, London: New York University Press.

Barbalet, Jack. 2002. Introduction: Why Emotions Are Crucial. In *Emotions and Sociology*. Oxford: Blackwell.

Behringer, Wolfgang. 2010. In *A Cultural History of Climate*, ed. Patrick Camiller. Cambridge, Malden: Polity Press.

Bell, John. 1992. Justice and the Law. In *Justice. Interdisciplinary Perspectives*, ed. Klaus R. Scherer. Cambridge, UK: Cambridge University Press.

Bell, John, and Erik Schokkaert. 1992. Interdisciplinary Theory and Research on Justice. In *Justice. Interdisciplinary Perspectives*, ed. Klaus R. Scherer. Cambridge, UK: Cambridge University Press.

Bettin Lattes, Gianfranco. 2011. L'agire affettivo. le forme dell'amore nelle scienze sociali. *Societàmutamentopolitica* 2 (4): 6.

Blau, Peter M. 1964a. *Exchange and Power in Social Life*. New York, London, Sydney: John Wiley & Sons.

———. 1964b. Justice in Social Exchange. *Sociological Inquiry* 34 (2): 193–206.

Blumer, Herbert. 1969. *Symbolic Interactionism. Perspective and Method*. Englewood Cliffs: Prentice-Hall.

Boltanski, Luc. 1990. *L'amour et La Justice Comme Compétences. Trois Essais de Sociologie de L'action*. Paris: Métailié.

Boncinelli, Edoardo. 2009. *Perché non possiamo non dirci darwinisti*. Milano: Rizzoli.

Boudon, Raymond. 1969. Relazioni tra proprietà individuali e proprietà collettive. In *Proprietà individuali e proprietà collettive: un problema di analisi ecologica*, ed. Raymond Boudon and Paul F. Lazarsfeld. Bologna: Il Mulino.

———. 1987. The Individualistic Tradition in Sociology. In *The Micro-Macro Link*, ed. Jeffrey C. Alexander, Bernhard Giesen, Richard Munch, and Neil J. Smelser. Berkeley, Los Angeles, London: University of California Press.

———. 2009. *Effets pervers et ordre sociale*. Paris: Quadrige - PUF.

Bourdieu, Pierre. 1981. Men and Machines. In *Advances in Social Theory and Methodology. Toward an Integration of Micro- and Macro-Sociologies*, ed. Karin D. Knorr-Cetina and Aaron V. Cicourel. Boston, London, Henley: Routledge and Kegan Paul.

Bovone, Laura. 1994. Goffman: micro o macro? In *Il rapporto micro-macro nelle teorie sociologiche contemporanee*, ed. Nicolò Addario. Milano: FrancoAngeli.

Campanale, Anna Maria. 2007. Introduzione. In *Bounded rationality e razionalità giuridica*, ed. Anna Maria Campanale and Her Simon. Bari: Adriatica Editrice.

Ciacci, Margherita. 1983. Significato e interazione: dal behaviorismo sociale all'interazionismo simbolico. In *Interazionismo simbolico*. Bologna: Il Mulino.

Cicourel, Aaron V. 1974. *Cognitive Sociology. Language and Meaning in Social Interaction*. New York: The Free Press.

Cohen, Michael D., James G. March, and Johan P. Olsen. 1972. A Garbage Can Model of Organizational Choice. *Administrative Science Quarterly* 17 (1): 1–25.

Collins, Randall. 1975. *Conflict Sociology. Toward an Explanatory Science*. New York, San Francisco, London: Academic Press.

———. 1981. Micro-Translation as a Theory Building Strategy. In *Advances in Social Theory and Methodology. Toward an Integration of Micro- and Macro-Sociologies*, ed. Karin D. Knorr-Cetina and Aaron V. Cicourel. Boston, London, Henley: Routledge and Kegan Paul.

———. 1987. Interaction Ritual Chains, Power and Property: The Micro-Macro Connection as an Empirically Based Theoretical Problem. In *The Micro-Macro Link*, ed. Jeffrey C. Alexander, Bernhard Giesen, Richard Munch, and Neil J. Smelser. Berkeley, Los Angeles, London: University of California Press.

———. 1988. *Theoretical Sociology*. San Diego: Harcourt Brace Jovanovich.

———. 2004. *Interaction Ritual Chains*. Princeton, Oxford: Princeton University Press.

Cortellazzi, Silvia, and Italo Piccoli. 1988. Nella prospettiva micro: la proposta etnometodologica. In *Sociologie micro, sociologie macro*, ed. Laura Bovone and Giancarlo Rovati. Milano: Vita e pensiero.

Crespi, Franco. 1989. *Azione sociale e potere*. Bologna: Il Mulino.

Crozier, Michel. 2010. *The Bureaucratic Phenomenon*. New Brunswick, NJ, London: Transaction Publishers.

Cullen, Bernard. 1992. Philosophical Theories of Justice. In *Justice. Interdisciplinary Perspectives*, ed. Klaus R. Scherer. Cambridge, UK: Cambridge University Press.

Dal Lago, Alessandro. 1994. L'etnometodologia e il problema dei rapporti micro-macro. In *Il rapporto micro-macro nelle teorie sociologiche contemporanee*, ed. Nicolò Addario. Milano: FrancoAngeli.

Denzin, Norman K. 1983. Interazionismo simbolico ed etnometodologia. In *Interazionismo simbolico*. Bologna: Il Mulino.

Di Lucia, Paolo, and Lorenzo Passerini Glazel. 2015. Norma forma: Hans Kelsen oltre il formalismo. In *Che cos'è la giustizia?* ed. Hans Kelsen. Macerata: Quodlibet.

Donati, Pierpaolo. 1983. *Introduzione alla sociologia relazionale*. Milano: FrancoAngeli.

Durkheim, Émile. 1912. *The Elementary Forms of Religious Life. [2001]*. Oxford: Oxford University Press.

———. 1917. In *The Rules of Sociological Method. [1982]*, ed. W.D. Halls. New York, London, Toronto, Sydney: The Free Press.

Eckhoff, Torstein. 1974. *Justice. Its Determinants in Social Interaction*. Rotterdam: Rotterdam University Press.

Elias, Norbert. 2000. *The Civilizing Process. Sociogenetic and Psycogenetic Investigations*. Malden, MA: Blackwell.

Elwell, Frank W. 2013. *Sociocultural Systems: Principles of Structure and Change*. Edmonton: Athabasca University Press.

Emirbayer, Mustafa. 1997. Manifesto for a Relational Sociology. *American Journal of Sociology* 103 (2): 281–317.

Ferrari, Vincenzo. 1995. *Giustizia e diritti umani. Osservazioni sociologico-giuridiche*. Milano: FrancoAngeli.

Fitzi, Gregor. 2011. Il contributo weberiano alla sociologia delle emozioni. *Societàmutamentopolitica* 2 (4): 37–50.

Frank, Jerome. 1949. *Law and the Modern Mind*. London: Stevens & Sons Limited.

Franks, David D. 2010. *Neurosociology. The Nexus between Neuroscience and Social Psychology*. New York: Springer.

French, Peter A. 2001. *The Virtues of Vengeance*. Lawrence: University Press of Kansas.

Gallino, Luciano. 1987. *L'attore sociale. biologia, cultura e intelligenza artificiale*. Torino: Giulio Einaudi.

Garfinkel, Harold. 1967. *Studies in Ethnomethodology*. Englewood Cliffs: Prentice-Hall.

Gerstein, Dean R. 1987. To Unpack Micro and Micro: Link Small with Large and Part with Hole. In *The Micro-Macro Link*, ed. Jeffrey C. Alexander, Bernhard Giesen, Richard Munch, and Neil J. Smelser. Berkeley, Los Angeles, London: University of California Press.

Girard, René. 1986. *The Scapegoat*. Baltimore: The Johns Hopkins University Press.

Goffman, Erving. 1959. *The Presentation of Self in Everyday Life*. Garden City: Doubleday & Co.

———. 1983. Rappresentazioni fuorvianti. In *Interazionismo simbolico*. Bologna: Il Mulino.

Greenberg, Jerald, and Ronald L. Cohen. 1982. Why Justice? Normative and Instrumental Interpretations. In *Equity and Justice in Social Behavior*, ed. Jerald Greenberg and Ronald L. Cohen. New Haven, London: Academic Press.

Habermas, Jürgen. 1986. *Teoria dell'agire comunicativo. 1. Razionalità nell'azione e razionalizzazione sociale*. Bologna: Il Mulino.

Harré, Rom. 1992. In *La costruzione sociale delle emozioni*, ed. Ornella Pavirani Arcangeli. Milano: Giuffrè.

Homans, George C. 1961. *Social Behavior: Its Elementary Forms*. New York, Burlingame: Harcourt, Brace & World.

Hume, David. 1739. In *A Treatise of Human Nature. [1896]*, ed. L.A. Selby-Bigge. Oxford: Clarendon Press https://www.files.ethz.ch/isn/125487/5010_Hume_Treatise_Human_Nature.pdf.

Huxley, Thomas Henry. 1989. In *Evolution and Ethics*, ed. James Paradis and George C. Williams. Princeton: Princeton University Press.

Iagulli, Paolo. 2011. *La sociologia delle emozioni*. Milano: FrancoAngeli.

Iorio, Gennaro. 2011. Agape: un concetto per le scienze sociali. *Societàmutamentopolitica* 2 (4): 101–114.

Jacoby, Susan. 1983. *Wild Justice: The Evolution of Revenge*. New York: Harper & Row.

Kahan, Dan M. 1999. The Progressive Appropriation of Disgust. In *The Passions of Law*, ed. Susan Bandes. New York, London: New York University Press.
Knorr-Cetina, Karin D. 1981. Introduction: The Micro-Sociological Challenge of Macro-Sociology: Towards a Reconstruction of Social Theory and Methodology. In *Advances in Social Theory and Methodology. Toward an Integration of Micro- and Macro-Sociologies*, ed. Karin D. Knorr-Cetina and Aaron V. Cicourel. Boston, London, Henley: Routledge and Kegan Paul.
Kurzweil, Edith. 1987. Psychoanalysis as the Macro-Micro Link. In *The Micro-Macro Link*, ed. Jeffrey C. Alexander, Bernhard Giesen, Richard Munch, and Neil J. Smelser. Berkeley, Los Angeles, London: University of California Press.
Lenski, Gerhard. 2005. *Ecological-Evolutionary Theory. Principles and Applications*. Boulder, London: Paradigm Publishers.
Lorini, Giuseppe, and Michelina Masia, eds. 2015. *Antropologia Della Vendetta*. Napoli: Edizioni scientifiche italiane.
Luhmann, Niklas. 1978. *Sistema giuridico e dogmatica giuridica*. Italian translation by Alberto Febbrajo. Bologna: Il Mulino.
Malinowski, Bronislaw. 1939. The Group and the Individual in Functional Analysis. *American Journal of Sociology* 44 (6): 938–964.
Markovsky, Barry. 1991. Prospects for a Cognitive-Structural Justice Theory. In *Social Justice in Human Relations: Societal and Psychological Origins of Justice. 2*, vol. 50. New York: Plenum Press.
Massaro, Toni M. 1999. Show (Some) Emotions. In *The Passions of Law*, ed. Susan Bandes. New York, London: New York University Press.
McCarthy, E. Doyle. 1989. Emotions Are Social Things. An Essay in the Sociology of Emotions. In *The Sociology of Emotions. Original Essays and Research Papers*, ed. David D. Franks and E. Doyle McCarthy. Greenwich CT, London: Jai Press.
———. 1994. The Social Construction of Emotions: New Directions from Culture Theory. In *Social Perspectives on Emotion*, ed. William Wentworth and Darrell Yardley. Bingley: Emerald.
Mead, George Herbert. 1972. In *Mind, Self & Society: From the Standpoint of a Social Behaviorist*, ed. Charles W. Morris. Chicago: University of Chicago Press.
Merton, Robert King. 1968. *Social Theory and Social Structure*. New York: The Free Press.
Meyering, Theo C. 2000. Physicalism and Downward Causation in Psychology and the Special Sciences. *Inquiry* 43 (2): 181–202.
Morin, Edgar. 1977. *La méthode. 1. La nature de la nature*. Paris: Éditions du Seuil.
Mutti, Antonio. 1994. La teoria della scelta razionale: dalla microriduzione alla soluzione combinatoria. In *Il rapporto micro-macro nelle teorie sociologiche contemporanee*, ed. Nicolò Addario. Milano: FrancoAngeli.
Nagel, Thomas. 1974. What Is It Like to Be a Bat? *The Philosophical Review* 83 (4): 435–450.
Niskanen, William A. 1971. *Bureaucracy and Representative Government*. Chicago, IL: Aldine-Atherton.
Nussbaum, Martha C. 1999. 'Secret Sewers of Vice'. Disgust, Bodies and the Law. In *The Passions of Law*, ed. Susan Bandes. New York, London: New York University Press.
———. 2004. *Hiding from Humanity. Disgust, Shame, and the Law*. Princeton, Oxford: Princeton University Press.
Pannarale, Luigi. 2008. *Il diritto che guarda. Rischi della decisione giuridica*. Milano: FrancoAngeli.
Pareto, Vilfredo. 1980. In *Scritti sociologici minori*, ed. Giovanni Busino. Torino: UTET.
———. 1981a. In *Trattato di sociologia generale. 1. Azioni Non -logiche. teorie filosofiche e pseudo-scientifiche*, ed. Pietro Rossi. Torino: Edizioni di Comunità.
———. 1981b. In *Trattato di sociologia generale. 2. I residui*, ed. Pietro Rossi. Torino: Edizioni di Comunità.
———. 1981c. In *Trattato di sociologia generale. 3. Le derivazioni*, ed. Pietro Rossi. Torino: Edizioni di Comunità.

Parsons, Talcott. 1964. *Social Structure and Personality*. New York: Free Press of Glencoe.
Passeron, Jean-Claude. 2006. *Le raisonnement sociologique*. Paris: Éditions Albin Michel.
Perelman, Chaïm. 1959. *La giustizia*. Italian translation by Liliana Ribet. Torino: Giappichelli.
Pigliaru, Antonio. 1959. *La vendetta barbaricina come ordinamento giuridico*. Milano: Giuffrè.
Popper, Karl Raimund. 1971. *The Open Society and Its Enemies. 2. The High Tide of Prophecy: Hegel, Marx and the Aftermath*. Princeton: Princeton University Press.
Posner, Richard A. 1999. Emotion Versus Emotionalism in Law. In *The Passions of Law*, ed. Susan Bandes. New York, London: New York University Press.
———. 2004. *Frontiers of Legal Theory*. Cambridge, MA, London: Harvard University Press.
Ravenna, Marcella. 2012. Stanley Milgram: le ricerche sulla sottomissione all'autorità. In *Ricerche e protagonisti della psicologia sociale*, ed. Augusto Palmonari and Nicoletta Cavazza. Bologna: Il Mulino.
Runciman, Walter G. 1989. Evolution in Sociology. In *Evolution and Its Influence*, ed. Alan Grafen. Oxford: Clarendon Press.
Scherer, Klaus R. 1992. Issues in the Study of Justice. In *Justice. Interdisciplinary Perspectives*, ed. Klaus R. Scherer. Cambridge, UK: Cambridge University Press.
Schütz, Alfred. 1967. *The Phenomenology of the Social World*. Chicago: Northwestern University Press.
Searle, John R. 2001. *Rationality in Action*. Cambridge, MA: The MIT Press.
Shott, Susan. 1979. Emotion and Social Life: A Symbolic Interactionist Analysis. *American Journal of Sociology* 84 (6): 1317–1334.
Simmel, Georg. 1955. In *Conflict (Streit)*, ed. Kurt H. Wolff. New York: The Free Press.
———. 1982. In *La Differenziazione Sociale*, ed. Bruno Accarino. Bari: Laterza.
———. 1984. *On Women, Sexuality and Love*. New Haven and London: Yale University Press.
Simon, Herbert A. 1961. *Administrative Behavior*. New York: Macmillan Press.
———. 2007. In *Bounded rationality e razionalità giuridica*, ed. Anna Maria Campanale. Bari: Adriatica Editrice.
Simonelli, Maria Ausilia. 2006. *Diritto e sentimenti. Il pensiero giuridico di Vilfredo Pareto*. Napoli: Edizioni scientifiche italiane.
Sims Bainbridge, William. 1997. The Omicron Point. Sociological Application of the Anthropic Theory. In *Chaos, Complexity, and Sociology. Myths, Models, and Theories*, ed. Raymond A. Eve, Sarah Horsfall, and Mary E. Lee. Thousand Oaks, London, New Dehli: Sage.
Smith, Adam. 2002. In *The Theory of Moral Sentiments*, ed. Knud Haakonssen. Cambridge: Cambridge University Press.
Solomon, Robert C. 1999. Justice v. Vengeance. On Law and the Satisfaction of Emotion. In *The Passions of Law*, ed. Susan Bandes. New York, London: New York University Press.
Sorokin, Pitrim A. 1950. *Altruistic Love. A Study of American Good Neighbors and Christian Saints*. Boston: The Beacon Press.
Stone, Julius. 1966. *Law and the Social Sciences in the Second Half Century*. Minneapolis: University of Minnesota Press.
Strahlendorf, Peter. 1992. Traditional Legal Concepts from an Evolutionary Perspective. In *The Sense of Justice. Biological Foundations of Law*, ed. Roger D. Masters and Margaret Gruter. Newbury Park, London, New Dehli: Sage.
Talhelm, T., X. Zhang, S. Oishi, C. Shimin, D. Duan, X. Lan, and S. Kitayama. 2014. Large-Scale Psychological Differences Within China Explained by Rice Versus Wheat Agriculture. *Science (New York, N.Y.)* 344 (6184): 603–608.
Tamanaha, Brian Z. 1997. *Realistic Socio-Legal Theory. Pragmatism and a Social Theory of Law*. Oxford: Clarendon Press.
Tammelo, Ilmar. 1957. Law, Justice, and Social Reality. *Osterr. Zeitschrift Fur Offentliches Recht* 8 (3): 383.
Terradas Saborit, Ignasi. 2008. *Justicia vindicatoria. De la ofensa e indefensión a la imprecación y el oráculo, la vindicta y el talión, la ordalía y el juramento, la composición y la reconciliación*. Madrid: Consejo Superior de Investigaciones Científicas.

Thoits, Peggy A. 1995. Devianza emozionale: futuri obiettivi della ricerca. In *Sociologia delle emozioni*, ed. Gabriella Turnaturi. Milano: Anabasi.

Tocqueville, Alexis de. 1840. *Democracy in America*. Kindle Edition. http://www.amazon.it/dp/B0082ZJNN0/ref=pe_364731_36330211_M1T1DP.

Turnaturi, Gabriella, ed. 1995. *La sociologia delle emozioni*. Milano: Anabasi.

———. 1998. La sociologia delle emozioni. In *Le nuove frontiere della sociologia*, ed. Paolo De Nardis. Roma: Carocci.

von Hayek, Friedrich August. 2013. *Law, Legislation and Liberty: A New Statement of the Liberal Principles of Justice and Political Economy*. London, New York: Routledge.

Wallerstein, Immanuel, Calestous Juma, Evelyn Fox Keller, Jurgen Kocka, Y.V. Mudimbe, Kinhide Mushakoji, Ilya Prigogine, Peter J. Taylor, and Michel-Rolph Trouillot. 1996. *Open the Social Sciences. Report of the Gubelkian Commission on the Restructuring of the Social Sciences*. Stanford: Stanford University Press.

Watson, Alan. 1985. *The Digest of Justinian*. Vol. 1. Philadelphia: University of Pennsylvania Press.

Weber, Max. 1978. In *Economy and Society. An Outline of Interpretive Sociology*, ed. Guenther Roth and Claus Wittich. Berkeley, Los Angeles, London: University of California Press.

———. 1994. In *Political Writings*, ed. Peter Lassman and Ronald Speirs. Cambridge: Cambridge University Press.

Wentworth, William M., and John Ryan, eds. 1994. *Social Perspectives on Emotion*. Bingley: Emerald.

Wentworth, William, and Darrell Yardley. 1994. Introduction. In *Social Perspectives on Emotion*, ed. William Wentworth and Darrell Yardley. Bingley: Emerald.

Williams, Simon. 2001. *Emotion and Social Theory*. Thousand Oaks, London, New Dehli: Sage.

Wippler, Reinhard, and Siegwart Lindenberg. 1987. Collective Phenomena and Rational Choice. In *The Micro-Macro Link*, ed. Jeffrey C. Alexander, Bernhard Giesen, Richard Munch, and Neil J. Smelser. Berkeley, Los Angeles, London: University of California Press.

Chapter 2
The Psychic Subject as a Legal Actor

2.1 Law and Psychology

Law and psychology are disciplines that live in separate wings of the same palace. Both share the goal of predicting and controlling human behavior. However, the law relies necessarily, though often unknowingly or naively, on psychological theories and paradigms it cannot do without. With regard to psychological paradigms, *sociological creationism* has implicitly expressed the same indifference to which interactionism and other microsociological theories had been condemned, thus repeating the old mistake of placing psychology and sociology on two separate levels that do not communicate. The gist of this critique sounds more or less like this: psychology studies behavioral processes that are inevitably individual and is thus too "contaminated" by the biological factor to offer a satisfactory narrative of social processes.

In this chapter, I will argue that if the cognitive sociology of law is to arrive at valid hypotheses about individual legal attitudes, it cannot do without those psychological paradigms that are relevant to explaining legal attitudes and cognition. As we are reminded of in studies that admirably reconstruct the relations between law and psychology (Bona and Rumiati 2013; Guthrie et al. 2001; Rachlinski 2000a, b), while the law relies on a historical record of human behavior that is idiographic—informed by ideology, anecdotes, and historical accident—psychology starts from a quite different set of assumptions, for they are essentially empirical and nomothetic. This view of psychology as an essentially nomothetic investigation directly informs Canter's (2008, p. 3) approach to Law & Psychology. In a similar vein, almost two decades ago Rachlinski (2000b, p. 766) commented that psychology, having been primed on issues already "pre-digested" by the economic analysis of law, was just then beginning to yield significant results in its application to law, predicting that the best results were yet to come. The areas of interaction between law and psychology are numerous and cover a wide range of human behaviors that have already become an object of research in the sociology of law and in social psychology. Brooks-Gordon and Freeman (2006, p. 4) have drawn up a preliminary list of issues

L. Cominelli, *Cognition of the Law*,
https://doi.org/10.1007/978-3-319-89348-8_2

that are affected by interaction: the attribution of motives; social identity; values beliefs, attitudes, and their relation to behavior; persuasion and changes in attitude; compliance and social influence; group dynamics; leadership and decision-making; prejudice; inter-group behavior; aggression; cognition and language; and culture. Cross-contamination can also clearly be appreciated in the socio-psychological study of law. As we will see, the line between social psychology and interactionist sociology is often very thin.

Even when no explicit acknowledgment is made, legal scholars and practitioners intuitively perceive the psychological relevance of the law. The extent of this relevance has been widely illustrated by Justickis (2008, pp. 224–225). In the first place, the impact that legal rules have on human behavior can only be reconstructed through a chain of psychological events, since legal regulation must be known, understood, stored in memory, and recalled at the right time, and the memory must be powerful enough to excite fear or emotion and thus prompt an act of compliance or violation. In the second place, the law, in any of its areas, inevitably affects every individual's personality. In the third place, any kind of legal regulation must meet minimum psychological requirements if it is to be (psychologically) perceived as "legitimate" (fair and worthy of enforcement).

The main channel through which psychological studies begin to be accepted in legal scholarship is through the *psychology of law* (Gulotta and Curci 2010; Quadrio and De Leo 1995). Indeed, it was as far back as the 1920s that, in Italian judicial practice, psychologists first began to be heard as expert witnesses in the courtroom. Psychology as applied to law is classified by Ferri into four areas of study: (1) *criminal psychology*, focused on the criminal's profile as perpetrator of a crime; (2) *judicial psychology*, centered on the criminal's behavior as an accused; (3) *prison psychology*, which studies the convicted criminal in the imprisonment stage; and (4) *legal psychology*, which coordinates psychological and psychopathological notions when it is necessary to evaluate particular conditions affecting a determination of criminal responsibility and punishment, such as age, drunkenness, or other transient mental states affecting the assessment of criminal behavior (Ferri 1925, p. xxiii).

Subsequently, Quadrio and De Leo suggested a newer breakdown of the different areas of psychology applied to law, identifying a first area in *legal psychology* proper; a second area in the *psychological study of organization activities*, focused in particular on the dynamics and strategies of the judiciary; a third area in the *psychology of sentencing*, concerned with the enforcement of judicial decisions and the treatment to which the law subjects those who have been found to be in violation (this area includes a subarea devoted to the psychology involved in the training of judicial staff and personnel); a fourth area in the *psychology of developmental risk*, concerned with situations that may put minors in danger, and specializing in the protection of juveniles; and finally a fifth area in the *psychology of deviance* proper, which studies deviant behaviors that may have criminal consequences.

Equally convincing, albeit less articulated, is the following tripartite structure of the relations between psychology and law (Berti 2002, pp. 24–25). We have "psychology *in* law," where psychology is understood to relate *functionally* to law, and in which it is the latter that dominates and introduces the questions it deems relevant, as in the case of expert inquiries requested in criminal proceedings to determine whether the defendant can be found guilty as charged (imputability) or whether a witness is credible. Then we have "psychology *and* law," where psychology is entrusted with a critical role of equal standing with law, investigating, for example, the human and subjective factor in judicial or negotiating decisions. And, finally, we have a "psychology *of* law," where the dominant role, from a methodological point of view, is played by psychology, entrusted with investigating our perception of legal norms and individual motives behind the framing norms, the drive to violate or respect them, and the intuitive criteria that guide the attribution of responsibility.

In an open and cognitive vision of legal attitudes, law and psychology should be much more closely interconnected than they are. This lack of interconnection is explicitly identified as a gap to be closed (Canter and Žukauskienė 2008). It has been remarked, not without reason, that law is defensively closed to other social sciences, because it prefers stability to efficiency (Justickis 2008, p. 234), even if a simplified and semi-autonomous "legal" psychology has been built, which could prove to be more than sufficient to this end (ibid., pp. 228–230). With a few exceptions, however, philosophical and sociological analyses have never overcome their ill-concealed distrust of psychology. To quote a contemporary philosopher of law, the "traditional philosophical categories are obsolete and stand as roadblocks barring the way to a more adequate understanding of the internal psychological aspect of individual and social human phenomena" (Pattaro 2005, p. 403). This is equally true of the sociology of law, testifying to the existence of an anti-psychological, anti-individualist bias working away in the background, with the short season of interactionist microsociology representing the exception that proves the rule.

To elaborate on the exceptions just mentioned, Petrażycki (2011) argued for a psychological view of the law according to which legal phenomena occur in the psyche. Without going into an exegesis of the Polish sociologist's work, and relying on the contemporary analysis of his work, his view is that the science of law deals essentially with illusions. Rules are the content of individual ethical judgments, even though Petrażycki did not explain what exactly is meant by "content" (Fittipaldi 2012, p. 31). One of the most relevant insights in Petrażycki's theory lies in his way of framing the distinction between the moral phenomenon and the legal phenomenon. But we have a definite difference from legal realism, according to which law aims to elicit or quell an emotion: the position between the two elements is inverted here, because it is precisely emotion that brings forth the illusion that takes the name of law. "Law—in the sense of a special class of real phenomena—is to be understood as those ethical experiences whose impulsions have an attributive character. All other ethical experiences—those connected with purely imperative impulsions—we shall call moral phenomena" (Petrażycki 2011, p. 61). Petrażycki's conception is socio-psychological in its fullest sense, though heterogeneous relative to the canons of sociology *and* psychology. If it weren't for the logical inversion just pointed out, it would be indistinguishable from positions which are in substance of legal realism, and according to which "a norm is [...] a motive of behaviour: It is the belief (*opinio vinculi*) that a certain type of action must be performed, in the normative sense of this word, anytime a relevant type of circumstance gets validly instantiated" (Pattaro 2005, p. 97). This motive and belief is received by the external system (the macro side), but in order for it to be considered fully normative, it must be internalized by the human agent (the micro side) in his brain (Pattaro 2005, p. 359). Of particular interest are two insights developed by Petrażycki: his definition of *legal impulsions* (Petrażycki 2011, p. 49), and his account of an individualized intuitive law depending on individual characteristics (ibid., p. 225).

Of a quite different nature is the analysis proffered by Amsterdam and Bruner (2000). They build on paradigms of cognitive psychology to construct not just a theory of law properly so called, but what they term a *legal pedagogy* (ibid., p. 289) based on the natural and individual tendency to categorize and on the centrality of narratives in the life of the law: "Law is not simply a system of ideas but a series of consequences that human beings inscribe on the lives of other human beings through the medium of those ideas" (ibid., p. 6). Categorization as an innate tendency that individuals display toward the law calls for the use of conceptual patterns and mental maps that have become indispensable to modern cognitive science. Categories are ubiquitous and unavoidable in mental phenomena (ibid., p. 19). Categories play a role comparable to cognitive heuristics, making possible mental resource-saving, instant practical utility, simplified comparison with peer groups, personal gratification, and risk control (ibid., pp. 21–25).

To this heuristic categorization Amsterdam and Bruner ascribe two main cultural functions: one is to achieve cohesion within a cultural group and the other is to impose the dominant categorial system on other groups (ibid., p. 34). They also point out some rules that govern the use of categories. Categories are culturally constructed and structure meaning by postulating a world that contains them. Categories necessarily have a function and become inextricable from praxis, even though they are never definitive (ibid., pp. 27–37). Each person finds it instinctive and natural categorizing people's behavior as intentional rather than as circumstantial or structurally determined (*attribution bias*). Categorization is based on the use of scripts, filled with familiar characters who have predictable behaviors in recurring situations, as well as in narratives, that is, complex schemes where the script is violated owing to obstacles or dangers, and which are generally described as tragedy, comedy, romance, or irony (ibid., pp. 45–46).

Narratives work on another level relative to theory: they do not contradict theory but cross it "longitudinally." There are alternative theories by which to explain the human ability and tendency to elaborate and pass on stories. This capacity can more easily be acquired by individuals whose language makes it possible to describe deviations from the ordinary: this replicates and extends one's mimetic abilities, thus making it possible to rebuild the world according to one's own categories, and giving a reproductive advantage to an individual with this talent (ibid., p. 116).

In Bruner and Amsterdam's pedagogy, narration is essential in the legal world "for reasons both banal and deep" (ibid., p. 110). Many legal categories derive from narrative (ibid., p. 12). Law is adversarial, and legal statements, except the trivial ones, are interpretive (ibid., pp. 6–7). Narration, with its load of characters and plots, whether fictional or nonfictional, enriches the expressive capacity of legal proceedings and protects us from their dangers (ibid., p. 113). Narrative is essential in litigation or negotiation, as well as in advocacy. It makes it possible to establish continuity with a precedent, to explain a decision or imbue an outworn theory with new strength (ibid., p. 140). Narrative ordinarily takes into account the existence of a "normative" element, understood as a normal or ordinary state of affairs which human actions have somehow complicated, and which needs to find a solution (ibid., p. 141).

To be sure, Amsterdam and Bruner's idea is interesting, and it could be worked into a broader vision in which law becomes the object of a cross-sectional study in the social sciences. The idea could become especially interesting when updated through empirical research on the judicial process (Di Donato 2008). However, the idea is no more than an intellectual stimulus, because the authors, by their own admission, were not trying to make their theory predictive. The legal pedagogy of narrative provides an interesting vantage point from which to duly underscore how the innate tendency to categorize is fundamental in the process of legal cognition, as it may support a more encompassing theory of cognitive bias, a study that, as we will see in Chap. 4, is going to form the basis of a modern cognitive sociology of law.

Psychoanalysis, too, has been a pervasive presence in psycho-legal analysis. In several respects psychoanalysis is relevant to the psychological study of law (Oneroso Di Lisa 1989). Interesting works underscore the centrality of psychoanalysis in legal behavior and attitudes by discussing the father figure (Frank 1949, p. 15) and the superego (Fittipaldi 2012, p. 33; Reis 1984, p. 32). In his work on psychoanalytic jurisprudence, Ehrenzweig reconstructs Freud's work in relation to the social sciences and in particular to the legal sciences. According to Ehrenzweig (1971), the relevance of psychoanalysis is not limited to the psycho-legal problem of the "life of the law" but also applies to the validity of norms as such. However, I believe that psychoanalytic theories can only be of marginal use in a cognitive socio-legal analysis. Although the question of the individual drive seems relevant to legal analysis, I wonder how useful it is to draw the distinction between Oedipal and post-Oedipal crimes, a distinction that hinges on whether or not a given criminal act was driven by an instinct that can be traced back to the time of a parricidal desire. How is this distinction relevant?

Ehrenzweig, for example, looks at research done studying the crimes reported in Denmark during the Nazi occupation. At that time, the Danish state police was dismantled, and there was a phase in which Danish criminal law was not being applied. What the data seemed to show was that in the absence of external coercion, the number of the so-called "Oedipal crimes" had only minimally changed, while post-Oedipal crimes saw a remarkable increase. The latter could therefore be considered the only crimes on which legal punishment actually had a deterrent effect (Ehrenzweig 1971, pp. 213–215). A parallel may even be drawn between criminal law—which could pretend to pursue rational purposes such as deterrence and re-education, while in fact seeking irrational punishment—and civil law, which apparently responds to a need for rational redress but in fact is also partly based on the same irrational desire to satisfy an aggressive urge.

However, some of the more fruitful Freudian ideas, such as the Oedipus complex, penis envy, or the death drive, could ultimately amount to an erroneous reinterpretation of evolutionary instincts of adaptation. In adolescence, for example, a boy may have conflicts with his father: although the ostensible object of these conflicts may be the mother, they are more likely to have to do with her time and attentions, and they would not carry any sexual implications (Wright 1995, pp. 315–316). Daly and Wilson (1997, pp. 125–126) report that according to empirical research, parent-offspring competition is irrespective of the child's sex. I subscribe to the view that there is little predictive power that can be ascribed to psychodynamic theory or to any theory that sees human behavior as driven and motivated mainly by subconscious needs and impulses (Bandura 1986, p. 3).

2.2 The Social Psychology of Law

In the postwar period, psycho-sociological research began to provide a socially informed theory of law with elements relevant to an investigation of legal attitudes, this thanks to a few works that would go on to become famous and widely cited, such as those on authoritarian personality (Adorno et al. 1950); on conformism, obedience, and submission to authority (Milgram 1963); on the dependence of individual normative expectations on the career ambitions of peers, a phenomenon observed studying war veterans who had competed for a promotion in the army's administrative bureaucracy (Stouffer et al. 1949); and on the way that the political views of judges influence their decisions (Schubert 1965, 1974).

Social psychology is the field where sociology and psychology have assayed their potential for interdisciplinary discourse (Tesser and Schwarz 2001), and where remarkable contributions have been produced investigating the normativity of law. Social psychology is concerned with understanding and explaining how thoughts, feelings, and individual behaviors are influenced by the objective, imagined, or implicit presence of others (Hogg and Vaughan 2011, p. 4). Normative actions, attitudes to the law, and feelings of justice are variables that influence our perception of the other, and therefore make possible a full-fledged socio-psychological investigation. The need for safety, for example, is closely linked to one's attitude to the law and one's feeling of justice. Because all individuals have an innate need for safety, we look to the behavior of others as a benchmark against which to make social comparisons and decide between alternative courses of action. Empirical research in sociology and social psychology distinguishes several kinds of justice: as a system of norms, as an ideology, as a motive of behavior, and as a strategy of self-interest (Cook and Hegtvedt 1992, p. 190). I believe that in a reconstruction based on an individualistic and cognitive approach such as the one being put forward here, it will be useful to focus on the latter two.

From the research carried out in social psychology, the cognitive sociology of law has acquired in the first place the concept of *attitude*. This is an essential tool because it makes it possible to distinguish with a greater degree of granularity some elements of behavior that lie between intentionality and reflex action. An attitude is "a psychological tendency that is expressed by evaluating a particular entity with some degree of favor or disfavor" (Bohner and Schwarz 2001, p. 413). The primacy accorded to the concept of attitude over that of instinct reflects the view that not everything is fixed and inevitable, even outside rational deliberate action (Burnstein and Branigan 2001, p. 3).

Understanding the dynamics by which attitudes change is essential to explaining the subjective management of social information, as well as the processes by which people are persuaded and behaviors directed. The concept of attitude has evolved in such a way as to outline a subjective element which no longer includes motivational and behavioral components without distinction, and which today can be regarded as limited to its evaluative component, essentially coinciding with a sort of degree of approval, or with the possibility that a person may appreciate something (Schwarz and Gerd 2001, p. 436).

In the previous chapter, I dealt with *linguistic* or *symbolic cognitive sociology*. A similar line of investigation has emerged in social psychology, namely, *social learning theory*, where the basic idea is that we know more than we can say (Wegner 1982, p. 77), and that among the systems of social evaluation and cognition is justice (ibid., p. 84). The causation model is interrelational: environmental, personal, and behavioral factors are all understood as "interacting determinants of each other" (Bandura 1986, p. XI). So conceived, social learning is to social psychology what symbolic interactionism is to sociology.

Socio-cognitive psychology is characterized by Bandura through the assertion that while psychic stimuli provide little predictive power, constructed symbolic codes precisely indicate "which modeled actions are mastered and how well they are retained over time" (Bandura 1986, p. 14). *Social cognitive theory* hypothesizes that human behavior can be explained on the basis of three mutually interacting variables: individual behaviors, cognitive factors, and environmental events. A person's nature depends on these fundamental capacities: symbolization, prediction, learning from the experiences of others, self-regulation, and self-reflection (ibid., p. 18ff.). The motivation to act can be broken down into two classes: biological incentives and cognitive incentives (ibid., p. 232).

Bandura also formulates numerous propositions on social sanctions. The first is that the most severe sanctions carry with them the risk of unanticipated negative consequences. Especially "if applied without clarifying explanation, [they] can create inhibitions that generalize to socially desirable patterns as well" (Bandura 1986, p. 265), because once these inhibitions and fears are created, they are not easily eliminated. In some cases, sanctions will prompt one to associate with people who encourage the very behaviors the sanctions target. It may frequently happen, however, that mild sanctions are more effective than severe ones in promoting self-control. People who are induced to comply on the basis of a mild sanction experience a cognitive dissonance caused by the incongruity between their compliance and an inadequate external cause (Bandura 1986, p. 268). A further explanation of cognitive dissonance has been offered by Sampson (1983, p. 141): since the mind rejects dissonance, it would seek to reduce the dissonance by unconsciously manipulating its perception of reality.

Bandura cites several empirical studies in support of this assertion. In a given condition, for example, it was found that building up police forces had no substantial effect on crime rates (Levine 1975). Other studies seemed to suggest that capital executions would lead to an increase in the homicide rate within a short time, as if capital punishment raised the threshold of acceptable violence (Bowers and Pierce 1980), thus supporting the thesis that the primary goal of the most punitive sanctions is retribution (Bandura 1986, p. 333).

These speculations, especially regarding capital punishment, seem to be borne out in recent empirical research. Zimring et al. (2010) compared murder rates between Singapore and Hong Kong in the last decades, finding that these follow similar patterns in the context of radically divergent capital punishment policies. In this particular case, even Bandura's hypothesis would have to be rejected, since Singapore should have witnessed an increase in the homicide rate, simply in virtue of a much more frequent recourse to capital punishment.

Winkielman et al. (2001, p. 98) have proposed three views by which to reframe the field of social cognition: on a "fundamentalist" view, the cognitive and reasoning skills used by an individual would be the same regardless of the social nature of the stimulus; on a "building-block view," the cognitive processes affected by social events are of a different and higher complexity than those used in nonsocial events; finally, on a "realist" view, the mental processes used in dealing with nonsocial objects are derived from the processes used in dealing with social objects. Social cognition thus presents itself as the general hypothesis from which other forms of cognition are derived.

The empirical research that has been devoted to defining psychosocial paradigms in light of which to analyze legal attitudes has always been remarkable: certainly, it has shown these paradigms to be more useful to this end than sociological paradigms. A particularly interesting example is psycho-physiology, which has provided interesting clues even as concerns normative social behavior, and which brings back the emotional side to the analysis. Psycho-physiology has focused on the effects of specific emotions triggered by legal values and legal processes (West-Newman 2005, p. 306), this by studying how social situations may alter physiological parameters, such as respiratory frequency or the heart rate when engaging in deception and cheating, or upon realizing that we are different from those around us, and hence deviant.

The anger that accompanies resentment is considered the crucial emotion that drives individuals to claim their rights (ibid., p. 319). Why are these physical parameters altered, as when we approach a police checkpoint, even if we know we are not at fault, or when we receive some official letter by a public body or authority, even before we know what its contents are? Without indulging in a sociological reductionism in the opposite direction, following which one could even argue that bodily manifestations are the epiphenomena of a social process, what we can say is that the results of psycho-physiology unfortunately are still inconclusive. There is little consistency between self-evaluations and independent measurements—with insufficient correlation between different parameters—and it has proved difficult to replicate many of these experiments. By rejecting the criticisms advanced by constructionism, which has described emotions as "social artifacts, culturally and historically variable in meaning, experience, and expression" (ibid., p. 311), it is possible to alternatively call for a "social neuroscience" by which to remedy these shortcomings.

On this view, we can welcome the idea that the emotions are learnable aptitudes and are useful in furthering certain collective socio-cultural functions (ibid., p. 312). Initially, it was social psychology that supported the assumption that the individual is in the first place a "coherence seeker" (Palmonari and Cavazza 2012, p. 15). The behavioral and cognitive approach has rapidly established itself, bringing with it the conviction that the individual can be more properly defined as a "thinking body" that processes information, but this individual can wind up excessively relying on intuition in the attempt to autonomously complete incomplete information, while underestimating some facts by overrating others (ibid., p. 18). It was bearing this cognitive limitation in mind that the seemingly contradictory concept of "unconscious social cognition" (Banaji et al. 2001, p. 136) was coined, arguing that a good part of social psychology consists in the "the study of processes that operate outside conscious awareness and intention" (ibid., p. 135).

A further problem that traditionally comes up in the study of legal attitudes—and an object of investigation in social psychology—is stereotyping. If we consider the functioning of the legal machinery, the social emergence of moral categories that turn into legal categories is rooted in the structuring that actually results from stereotyping. The main hypothesis of labeling theories (Becker 1963; Goffman 1959) is that the identification of a criminal is informed by stereotypical social stigmas (Ghezzi 1996). Even in this case, we observe the homologies between psychological and sociological paradigms, which start from considerably different assumptions but come to the same results. Like legal rules, social stereotypes contain prescriptive and descriptive elements (Biernat and Billings 2001, p. 259). As experiments on racial prejudice have shown, the unconscious stereotype is ubiquitous and almost unpreventable (Banaji et al. 2001, p. 142). The controllability of unconscious stereotypes has sparked a lively debate, as well as provided strong incentives to do empirical research, because stereotypes affect social interactions even when those who are affected are vigilant against being stereotyped and are strongly motivated to prevent that.

Although this research remains focused on the functioning of the nervous system in the presence of basic stimuli, Pavlov (1928) sketched some hypotheses on the most evolved nervous activities: in so doing he underscored the need to move beyond the distinction between the psychological and the physiological, thus anticipating the development of cognitive neuroscience. In fact, like other unconscious cognitive mechanisms, stereotyping operates in semi-concealment. But it is also doubtful that they are unconscious, and it could be that even if clearly originated in the environment, they become hardwired as conditioned reflexes of our behavior.

Let an example illustrate. Imagine using an escalator every day in your commute to and from work. You develop the skill and habit of tilting your weight forward ever so slightly as you step onto the escalator and backward as you step off, so as to counterbalance the accelerating and decelerating force your body is subjected to in these two maneuvers, and in this way you can keep from falling. But then one day the escalator is broken (an experience we should all be able to relate to): you step onto the escalator, as usual, but you lose your balance because your instinct kicks in and you reflexively adjust your weight even though there is no accelerating or decelerating force to counteract your adjustment. What this example illustrates is that, as much as we may strive to suppress this instinctive counterbalancing movement, we cannot do so, because this reaction has become hardwired, and every time we step onto a broken escalator, we risk losing our balance.

But if the existence of reflexively conditioned behaviors is accepted for some basic needs (hunger, pain, sexual arousal), the same does not apply in the social sciences for social actions that are cognitively conditioned. Sociology tends to exclude the possibility of semi-involuntary social behaviors that are determined on an individual basis. This is what happens in the case of responses to some strongly social situations, involving behaviors that also fall within the scope of the law. As we can gather from empirical evidence showing how respiration is affected in the presence of certain social stimuli or social prejudices, these mechanisms operate in complex and sophisticated social situations (Lyon 1994).

To elaborate on that last point, respiration depends on metabolic systems, and specific respiratory habits can be associated with emotional states through conditioning. Respiratory patterns, with their associated reactions, can be understood as ways of communicating at different levels of depth (Lyon 1994, p. 88). Although it would not be impossible, in such cases, to counterbalance the social instinctive reflex, it would take a rather involved and complex deconditioning effort. There are many more elements in common than there are differences between behavioral reflexes and behavioral prejudices of a social nature (such as racial prejudice, but also doubting the goodwill of the counterparty in litigation). Prejudices require a more elaborate cognitive activity that sometimes enables individuals of a more rational bent to activate what has been termed "cognitive system no. 2," that is, the acquired system that evolved animals have. This system 2 is more analytical, slower, and cognitively demanding, but less susceptible to the errors we can run into by relying on the instinctive heuristics of "cognitive system no. 1" (Evans 2008; Kahneman 2011).

However, there is no clear boundary in our use of the two systems: both are in the continuum between behaviors, nonsocial actions, and social actions. Fodor's (1983) idea of a modularity of the mind leads to an awareness that our cognitive architecture is formed by several coordinated but independent processing systems, some of which were tasked with solving the adaptation problems our ancestors encountered. Today, we have come to define four or perhaps five basic mental systems: intuitive physics, which enables us to understand inanimate objects and mechanic interactions; naive psychology, which captures actions that conscious agents direct at a certain end; spatial cognition, which captures places and geometric relationships; naive mathematics, which is based on sets and their numerical sorting ratios; and, finally, though this is still controversial, a cognitive system that allows us to represent the *us* in-group in opposition to the *them* out-group (Marraffa 2011, p. 174).

2.3 Distributive Justice

More than any other question investigated in social psychology, it is the sense of justice that strikes me as most crucial to an understanding of legal attitudes. While the concept of justice is so broad as to prove elusive (Törnblom 1992, p. 177), the *sense* of justice is more manageable: it may be defined as a psychic sentiment, an intuition of what is fair in a broad sense (not in the specific sense we will see shortly), and it tells us what we instinctively consider to be the ultimate purpose of law. In socio-psychological terms, Finkel (2000) suggests the existence of a "commonsense justice" that does not necessarily conform to the justice dispensed by the courts under the rules of the law. Even with regard to justice I believe that a purely culturalist vision, inextricably linking any conception of justice to the social and cultural context, is destined for failure. In fact, I agree with Homans (1961, pp. 383–384) when he cautions us against the anthropological obsession with the cultural uniqueness of particular societies: human nature is the only true cultural universal.

The first distinction we can make in the matter of justice is between distributive and procedural justice: by distributive justice is meant here the fairness of the final decision or allocation, while procedural justice is understood to refer to the fairness of the procedure by which this allocation is decided. Distributive justice has been the object of some of the most remarkable work in social psychology (Deutsch 1985; Folger 1984b; Homans 1961; Vermunt 2014). Homans (1961, p. 264) believes that distributive justice is realized for someone "when each of the various features of his investments and his activities, put into rank order in comparison with those of other men, fall in the same place in all the different rank orders," thereby obtaining a congruity of status coupled with a social certainty. Homans underlines that implicit in every social action are some rules of justice understood to hold true: "A man in an exchange relation with another will expect that the rewards of each man be proportional to his cost—the greater the rewards, the greater the costs—and that the net rewards, or profits, of each man be proportional to his investments—the greater the investments, the greater the profits" (ibid., p. 232).

Our sense of justice is violated, and a corresponding a feeling of injustice emerges, when we compare our own investments, rewards, and costs with those of others (Homans 1961, p. 248). The notion of proportionality between investments and rewards appears to be central, and the differences between persons arise mainly because there is no agreement on what legitimately constitutes an investment as opposed to a cost or a reward. People think that there should be a relationship between effort and reward, between input and outcome (Allen 1982, p. 187).

Homans sees three justifications for this proportionality: (1) a direct relationship that results from specific links between investment and profit; (2) experiences with the natural environment; and (3) previous experiences of exchange. Through such experiences and relationships, individuals form probabilistic expectations about when a similar relationship between investments and profits ought to recur in the future. These probabilities become normative expectations, because precedents tend to become expectations (Homans 1961, p. 73). An unjust reward that repeats itself will, in time, come to be considered fair. Homans seems to observe a pragmatic evolution whereby, as cultural norms evolve, so does our sense of justice. In fact, the psychological sense of injustice triggered by the disproportion between an object and the asking price may be attributed to a moral, noneconomic reason (Jhering 1913, pp. 174–175).

Decidedly more sophisticated than Homans's account of distributive justice, however, is Morton Deutsch's account. Deutsch recognizes that, when it comes to intuitions of justice, the issues are much more numerous and complex than a description in terms of costs, investments, and rewards, because they concern the nature of the good or harm, the roles of the subjects involved in the distribution of such good or harm, the style and timing of the distribution, and the underlying values (Deutsch 1985, p. 2). Deutsch's basic principles of distributive justice are the principle of equity, the principle of equality, and the principle of need (ibid., p. 45). These three principles are arrived at by a culling process that starts out by taking many different criteria of justice into account: equity can be evaluated by looking at individual skills and potential, effort, improving performance, the principle of reciprocity, or a minimum standard of protection; equality can be considered both in terms of opportunities and results; and need may be complemented by external social standards that vary according to the intrinsic value of the contribution or to the common good (ibid., p. 5).

Each of these divergent criteria of distributive justice, however, can be traced back to the three basic principles. The first and perhaps most encompassing is the principle of equity. Unlike equality and need, equity is fundamentally based on a relation of exchange, which can be understood as market exchange or an exchange made in light of socioeconomic ends (Emerson 1976). The use of one criterion rather than the other depends on how the relationship is set up: whether it is market-oriented or socially oriented. When market norms supplant social norms, the latter can only be re-established with some effort (Ariely 2009, p. 80). This observation is confirmed by Sampson (1983, p. 176), who describes the following dynamic: when two persons are in a relationship of commonality and understand this to be a friendly relationship rather than a business or impersonal relationship, and one of the two persons asks for something as a reward for a favor done, the "attraction" between the two decreases. Thus, treating a relationship of commonality in terms of exchange compromises the relationship. It can also be observed that forcing help on those who consider themselves to be our peers is perceived by them as antagonistic (Homans 1961, p. 252), and that in some situations, declining a gift and entering into an exchange relationship would be equivalent to a declaration of hostilities (ibid., p. 318).

As to the matter of choosing between criteria of distributive justice, these can be observed to change depending on the context of a relationship, but also with age. In a study of adolescents, for example, it was found that when the subjects were bound by an affective relationship (a family relationship or one of friendship), equality clearly prevailed. In a bureaucratized relational context, by contrast, the criteria used were mainly those of merit or need (Bosisio and Cammarata 2007, p. 155). Adolescent boys were inclined toward autonomy and the identification of objective criteria of justice, preferring individual responsibility; females of the same age were instead inclined toward caring for fellow beings and the balancing of criteria, while also identifying collective responsibilities and attributing considerable weight to social conditioning (ibid., p. 157).

Piaget deals with the growth of the feeling of justice in children, and he points out that the sense of equality develops only after a certain age, when it outstrips all the other drives, and then in a third phase it combines with other considerations of equity, thereby developing into a more sophisticated conception (Krebs 1982, p. 276). The process through which the sense of justice is acquired has also been investigated by Fikentscher (1999, pp. 127–128). In his work studying German children and adolescents he identifies some questions: can moral development and the sense of justice be learned only partly, and what are their predispositions? Are there certain phases in adolescence or childhood in which these notions are learned? And, if so, is learning active or passive, conscious or unconscious? In another study, it was found that before the age of six, children rely on a self-interest principle or on equality, and only from the age of six do they begin to rely on equity, while proportional equity does not develop before the age of thirteen, in parallel with the concept of logical-mathematical proportionality (Major and Deaux 1982, p. 55). From the age of six, children judge physical violence more seriously than theft, demonstrating that even subjects who are deemed immature have complex insights straddling different fields, and that it is quite early that they begin to develop a legal understanding (Robinson et al. 2007, pp. 1669, 1672).

Folger identifies three distributive attitudes: *equal treatment*, according to which everyone receives an equal quantity; *leveling*, with which each person is brought to the same level; and a *social minimum*, where no one can be allowed to fall below a certain pre-established threshold (Folger 1984a). Additional variants can be hypothesized within these patterns. According to the principle of a *ceiling*, for example, no one can exceed a certain threshold of resources (think of antitrust regulations). The principle of *range* instead establishes that in a fair distribution, the difference between those at the top and those at the bottom should never exceed a conventional reference value (Folger 1984a, p. 15).

McClintock and Keil (1982) posit a further psycho-sociological exchange-theory model that I am using to clarify what is meant by distributive justice. On this model, human actors tend to form and perpetuate those social relationships that offer an opportunity to obtain rewards, and that are profitable in the sense that the rewards obtained exceed the costs deriving from maintaining the relationship. The relationship is kept going when it is discovered to be mutually profitable. The actors within the relationship are driven to reciprocate the rewards to ensure the continuity of the exchange. The exchange is not limited to economic activities but can also have a social nature (think of the different logic of gifting or helping). In an exchange relationship, costs and rewards may or may not be distributed symmetrically among the actors involved, and the actor with more valuable resources is assumed to be able to exercise greater influence or power in the relationship. Finally, the model universally assumes that an understanding of institutional arrangements is not sufficient to describe the process of social exchange (McClintock and Keil 1982, pp. 354–355).

These arrangements only set certain (institutional) boundaries within which human interaction and exchange can take place.

Deutsch takes issue with theories that posit equitable exchange as a fundamental criterion of justice. The theory of equity certainly suffers from some methodological problems, such as vagueness and the lack of empirical support, and above all it faces the problem of putting an objective or interpersonal value to rewards and punishments, whose value is often only subjective. But it is his theoretical criticism that carries the true weight of his objections. In the first place, theories based on equity do not characterize the relationship between the parties as strategic, thereby ignoring how everyone negotiates to influence the thoughts and actions of others (Deutsch 1985, p. 25). In the second place, motivational and cognitive implications may be unclear, that is, we cannot be sure that everyone tends to be a self-interested and economically rational maximizer (ibid., p. 29).

It is also necessary to underline how the criteria of justice often diverge in their substance from those of fair exchange. Thus, in the family and other welfare institutions, the governing criterion used is that of need, while in friendship and other solidarity relationships, the governing criterion is that of equality. Although need and equality are close relatives and sometimes cannot be distinguished, they are necessarily divergent (ibid., p. 43). In competitive relationships, the uncontroversially accepted principle of justice is winner-take-all: the winner takes the entire prize (ibid., p. 30). A threefold distinction must therefore be made, says Deutsch, regarding cooperative relationships, in which different logics can be used depending on the long-term purpose of the relationship: in cooperative relationships whose primary goal is economic productivity, *equity* will prevail over equality or need, and it will act as the dominant criterion of distributive justice; in cooperative relations whose primary goal is to promote or maintain good social relations, *equality* will instead prevail; in cooperative relationships where individual fulfillment and welfare are paramount, the dominant principle will be that of *need* (Deutsch 1985, pp. 38–42).

The paradigms that rejects the idea of equitable exchange base the sense of justice on equality, or rather, they claim that our perception of injustice is mainly motivated by the absence of equality. From several analyses it emerges that what elicits the feeling of justice is the feeling of equality, or, more to the point, the indignation felt at inequality triggers a sense that justice that not been served. Thus, in the *aurea mediocritas* Durkheim (1893, p. 185) sees the path to happiness, noting that pleasure is not accompanied by states of consciousness that are too intense, and that inequality tends to activate envy, fomenting the sense of injustice.

In sociology, the question is further addressed by reference to the still-famous *American Soldier* (Stouffer et al. 1949), focused on the analysis of postwar organizational dynamics in the U.S. military apparatus. It was paradoxically pointed out, in that context, that in corps where promotions were more frequent (as in the U.S. air force), higher levels of collective frustration were detected than in corps where promotions were rarer (as in the U.S. military police). Stronger "social mobility" within the organization made for a general aspiration to obtain easier promotions. It is widely accepted that dissatisfaction and social revolutions emerge in periods of increasing expectation, rather than in times of black despair (Deutsch 1985, p. 54), yet Stouffer's analysis neglects to consider some essential variables, and the paradoxical results it seems to lead to may only be apparent. In the work done by Stouffer and his colleagues, levels of frustration were measured with insufficiently sophisticated tools, and what they saw could have simply have been a short-term frustration brought on by envy. The researchers did not go back to observe the long-term effects on these frustrated subjects, or whether this frustration had changed into a renewed ambition or determination.

As Boudon (2009, p. 154) correctly notes, it is not that the prospect of promotion is wholly unrelated to levels of satisfaction, but that these variables depend on mutual structures of interdependence. Caution is called for even when evaluating the hypothesis that posits the existence of a *hedonistic treadmill*, that is, an imaginary escalator of social ascent that can never be stopped, but this hypothesis does plausibly suggest that the feeling of justice should be relativized to the reference social group (the one used as a benchmark of comparison). This social dynamic is highlighted by Merton: every satisfaction is relative, since happiness lies precisely in having what others do not have (Merton 1968, p. 196), and when a society exceeds the threshold of mobility from a state of persistent immobility, irrational self-propagating expectations are generated (Elster 1985, p. 139).

At the core of every conception of justice, and so at the core of distributive justice, lies the notion of "comparative allotment" (Cohen and Greenberg 1982, p. 1). These analyses form the basis of what is now termed the theory of "relative deprivation" (Sampson 1983, p. 4). At the origins of the notion, in addition to Stouffer's idea of the reference social group, we have Runciman's (1966) study of class dynamics in Great Britain in the twentieth century. Runciman's main insight is that the reference group is not just used as a benchmark of comparison and evaluation but can also take on a normative function (Runciman 1966, pp. 10–12): in this way, relative deprivation can be used as a heuristic instrument without entailing any value judgment, as happens with the sense of injustice or envy. Envy is a complex behavior guided by the differences perceived in the reward system and in social position. Complexity is introduced by the presence of additional ingredients, such as admiration, submission, and obedience, and hence by the explicit recognition of difference (Herrmann-Pillath 1994, pp. 63–64).

The theory of equity and the theory of relative deprivation are both relevant in defining the problem of distributive justice, but they have been forged in different disciplines (the former in social psychology and behavioral economics, the latter in organizational and political sociology), and rarely have they been brought into relation. The theory of relative deprivation originated primarily out of the work of Stouffer et al. (1949), but it is with Merton's (1968) work that the concept is strengthened in the social sciences and finally becomes firmly established. Relative deprivation occurs when a subject, lacking something he wants, compares himself with someone in his group who instead owns that thing. This awareness induces resentment, anger, dissatisfaction, disappointment, and unhappiness.

Less evident is the perception of injustice in subjects who do not belong to the reference group, a perception that can be defined as "relative subordination." The sociality of this emotion can be appreciated when we instinctively envy the success of our peers. The empirical relevance of relative deprivation leads us to the conclusion that, unlike what is assumed in theories of exchange, our conception of distributive justice is less interested in the exchange value than it is in the symbolic import or status value (Crosby and Gonzalez-Intal 1982, p. 159). Sociological analysis agrees that in most cases the meaning of actions is not to acquire resources but to have recognition (Crespi 1989, p. 126). And Boudon (2009, p. 155) was in this sense right: it is excessively pessimistic to fear that industrial societies based on meritocratic competition will inevitably engender a generalized situation of negativity and envy.

No less important than research on the sense of justice has been the research devoted to *injustice* and to what individual perceptions and reactions it leads to. The effects of perception of injustice are considerable and have been explored empirically in psycho-sociological literature. Here it is worth mentioning the study cited by Deutsch (1985, p. 14) according to which a group of individuals who are overpaid relative to what their abilities and effort deserve: once they are made aware of this fact, they spontaneously make up for this injustice by increasing the quality or quantity of work, the explanation being that the information placed them in a situation where their ego was involved, and so they attempted to prove that they were competent and worthy.

If we turn now from the perspective of the unfairly privileged subject to that of the unjustly deprived, we cannot neglect to consider the social-interaction simulation that has been widely used in game theory in recent years and is known as the *ultimatum game* (Nowak et al. 2000; Proctor et al. 2013). In this game, a situation of clear imbalance is set up between two players. One subject is given assets that place her in a position of comparative wealth relative to a destitute subject who is given no assets at all, and unilaterally—without having had any previous social relationships with the destitute subject, and without being able to receive requests or petitions—she must decide how to share those assets with the destitute subject, whose only option is to either refuse or accept the offer as stated: if he refuses, he will lose what was offered by the well-off subject, but this would also strip the latter of all her assets; if he accepts, the assets would be shared according to the offer, no matter how unbalanced and unequal it may be. Because the well-off subject has not had any prior acquaintance with the destitute subject, the game excludes the possible effect of a bond of social solidarity forming between the two, and the results show that, on average, the well-off subjects tend to offer 40% of the assets allotted to them, while the destitute subjects tend to refuse offers below 30% of the total value of the assets.

These two percentage points would therefore seem to bookend the range of a generalized and instinctual threshold of justice: in an offer below the 30% threshold, the perception of injustice would trigger a response (refusal of the offer) so strong that the offeree is willing to do damage to himself in order to punish the selfish player who is taking advantage of the unilateral situation. This apparently self-defeating, if altruistic, punishment (Fehr and Gächter 2002) may be irrational in the short term, but it can be seen to be clearly rational in long term: we punish selfish subjects so that in the future they will interact with others according to a prosocial attitude that is more prone to sharing. This over-reactive mechanism is frequently also found in court, especially when family cases trigger highly conflictual dynamics: in order to punish a "selfish" spouse, many are willing to adopt extremely aggressive tactics that in the short term are expensive and risky, solely with the intent to inflict as much damage as possible.

Another clear example of this dynamic is offered by cases of medical negligence, where it has been highlighted that victims and their relatives, following a harmful event, are often determined to identify who is responsible and the extent of his or her guilt, even when this does not afford any particular advantages. One study (Dauer 2006, pp. 257, 264) has shown that when the medical staff offers a prompt apology, the subsequent settlement proposal is accepted in a much higher percentage (73 as against 52%). The impact that conciliatory procedures have on compensation claims in medical malpractice cases is particularly evident in the facilities with a hospital ombudsman (Houk and Edelstein 2008).

Closely connected, in terms of exchange and context, is the issue of prosocial behavior as a set of acts that benefit others, such as donating, sharing, or offering help in emergency situations (Krebs 1982, p. 261). Many hypotheses have been put forward on why we reciprocate the help received: because we are attracted to those who help us, or to comply with an inherent norm of reciprocity, to offer repayment, or to remedy an injustice (Krebs 1982, p. 291). The principle of reciprocity is a finer articulation of the principle of fairness, specifying it at the motivational level. Reciprocity makes social sense because every social subject is motivated to maintain a balanced relationship with others (Krebs 1982, p. 295). From a cognitive standpoint, injustice caused by iniquity is considered a special case of "dissonance": we perceive a discrepancy between our own individual inputs and outcomes and those of a reference subject or group.

Major and Deaux have attempted a systematization of the individual differences that intervene in some cognitive processes by altering the perception of equity in the relationship between input and output. These "cognitive problems" can affect our perception of inputs, our evaluation of performance, our attributions of performance, our expectations about remuneration for a result, and our perception of result (Major and Deaux 1982, pp. 65–66). To cite only a few of these distortions in our perception of inputs, women, for example, are likely to attribute less value to their time and work than men, while as concerns the question of what their performance is to be attributed to, they tend to explain their successes by pointing to external factors and their own defeats by pointing to internal failures, and as concerns their expectations of reward, they attribute a lower value to money and a higher value to social approval (ibid., pp. 66–69), thus bearing out the view that they tend to play a more gregarious role than men so as to ensure the social group's cohesion.

In the allocation of goods involving third parties, there are no differences between men and women when they do not expect any interaction with the recipients of those goods. In allocations made to their own advantage, women tend to award themselves less money, and consider this to be fair. When it comes to allocating goods to coworkers, women tend to be more generous with men and women alike, while men tend to be more generous only with female coworkers. Both sexes assume that women prefer a more equitable division than men, and that greater generosity is necessary to reward women (ibid., pp. 46–48).

In a slightly different sense, the hypothesis has been made, on a culturalist approach, that we have "intuitions of justice." These are understood as a socially learned instinct that would promote norms that make the group more efficient in its functioning. However, if these intuitions are socially learned, then it proves difficult to explain their limited range, for they ought to reflect the variety of individual life experiences, and thus be much more varied than they in fact are. In response to this line of objection, two kinds of variations have been proposed in the theory of social learning. The first goes by the name of "spontaneous social learning": a nonevolutionary type of learning according to which shared conceptions of justice are the most efficient for the group, because they allow productive social cooperation. However, the assumption that individuals, without further incentives, would prefer norms that are more efficient for the group rather than for themselves could hardly pass a stricter test (Robinson et al. 2007, pp. 1679–1680). The second variation is that of "cumulative social learning." Unlike the first version, it posits that the most efficient rules for the group are learned by accumulation over several generations (ibid., p. 1681).

Robinson and colleagues, however, also reject this rescue attempt, this on the basis of various considerations that partly reflect the criticisms already mentioned. First of all, the diversity of environmental and cultural situations should make for a much greater variation in intuitions of justice than what may ordinarily be found. What can be observed is, instead, a substantial uniformity among radically different cultural and demographic groups. The social-learning account is also difficult to reconcile with the substantial uniformity in children's visions of morality and justice, which, as we have seen, emerge rather uniformly from the time they are in preschool. And even without the insights we gain from evolutionary psychology, the hypothesis that intuitions of justice are socially learned would still be inconsistent because it would be incompatible with the intuitive nature of this kind of reasoning, which is based on rapid instinctual evaluations. However, I believe that this last argument should be ignored because it is potentially tautological. The evolutionary and unconscious origin of our notions of justice seems already sufficiently defensible.

Finally, the fundamentally intuitive nature of our conceptions of justice can be supported by what strikes me as a more compelling argument, which is that it is difficult, if not impossible, to articulate in words the principles underpinning the decisions we make in matters of justice. This strongly suggests that these principles cannot be socially learned through language (Robinson et al. 2007, pp. 1682–1686). In supporting these conclusions, we can take a further cue from the observation that the criminal system in each country is rather slow to evolve, at least at its core, and does not tolerate such significant divergences from the conception of justice that prevails in its own society. In terms of empirical evidence, Deutsch reports the results of experiments that have investigated individual preferences when having to choose among divergent principles of justice, where the principal aim of the experiments was to understand the effect of incentives to cooperate as against incentives for competition or individualism (Frank and Cook 1995).

One common finding in these experiments is that when individuals are shrouded in a veil of ignorance about how they will perform relative to other subjects, they have a strong preference for *proportionality* (equity) or *perfect equality*, while the *need* and *winner-take-all* principles are the least appreciated. These preferences changed in favor of equity and winner-take-all only for those who discovered that they were the most productive subjects: the change was slight but consistent across both sexes. Winner-take-all societies are those socio-economic systems, growing in number, where in order to attract and reward talent, a higher portion of the total collective wealth accrues to some specific productive sectors and in particular to those at the top. These hyper-competitive markets could paradoxically cause social dysfunctions by stimulating unproductive and wasteful investment and consumption (ibid., p. 1995).

By contrast, some significant differences began to be acknowledged with experiments that involved carrying out an extended and highly interdependent task requiring a high degree of cooperation. While the people selected for the group in which a principle of equality was applied cooperated and achieved good results, the people in the group where the winner took all were highly competitive, interfering with one another and ending up with a lower score. Finally, those in the condition of proportionality were more individualistic, without cooperating or interfering, and their result was at midrange (Deutsch 1985, pp. 153–154). The extent of variations, however, did not suggest that the effects of the different incentives on performance were of great significance. As for the attitudes the participants showed under the different principles of justice that were being applied, the winner-take-all principle clearly seemed to be associated with a sense of security, power, and toughness; a more conservative political outlook; and a sense of competitiveness that led participants to downplay the contribution of teammates.

Deutsch (1985, pp. 196–203) concludes by laying out five propositions of justice that bring into relation the incentives and criteria of distributive justice applied in concrete situations: (1) cooperative systems for distributing rewards have more favorable effects than competitive ones, not only on individual and group productivity but also on individual learning, on social relations, on self-esteem, on the attitude taken in carrying the tasks one is assigned, and on the sense of responsibility toward the other members of the group; (2) in conflict situations, the ability of the parties to find a stable and mutually satisfactory agreement improves in conditions that typically support cooperation, while it worsens in conditions that typically encourage competition; (3) there is no clear empirical evidence that the members of a group and individuals work more productively when they expect to be rewarded according to performance rather than under an egalitarian or need-based reward system; (4) the preference for socio-centric principles of distributive justice (such as egalitarianism and generosity) is associated with positive and solidarity-oriented social relations, while preference for principles centered on individuality (such as proportionality or equity) is associated with impersonal, result-oriented, or economic social relations;

and (5) sensitivity toward injustice can be increased by providing social support for its recognition and viable options to put a remedy to it.

As further confirmation of the above conclusions, it was found that, in the context of organizations, remuneration systems based on group performance led to better results than remuneration systems based on individual performance (Deutsch 1985, p. 227). There are five conditions that need to be met in order to effectively motivate a specific behavior: (1) the rewards must be of a certain amount; (2) they must be specifically gauged to different behaviors; (3) these behaviors must be measurable on the basis of objective criteria; (4) information must be provided that makes it clear how the rewards are allocated; and (5) there must be a high level of trust in the reward system and a strong acceptance of it. When these conditions are not met, a system based on rewarding performance can only contribute to generating distrust, lack of communication, falsification of data, and dysfunctional individual behavior in relation to the organization's objectives.

2.4 Procedural Justice

In the 1970s there started to develop a line of socio-psychological research that focused on the idea of *procedural justice*. At its basis was a hypothesis about the way individuals form their personal intuitions of justice, which is that people consider no less important than the final distribution of rewards and punishments, if not more important, the way in which they are treated and the legal decision-making process. The concept is aptly summarized in Stuart Hampshire's (2000, p. 4) remark that the fundamental kind of equity in conflict-resolution is the fairness of the procedure, a constant of human nature that is recognized as a value in most cultures and ages. This is a research line that finds fertile ground especially where the legal tradition of liberal democracies places a high value on procedure, but this does not mean that the same hypothesis cannot be generalized to other legal traditions. Significant empirical work on this subject has been carried out by Thibaut and Walker, a psychologist and a jurist (Thibaut et al. 1974; Thibaut and Walker 1975). They start from the Rawlsian concept of justice, or from principles of justice that individuals are willing to subscribe to on the reasoning that such principles will protect them from unforeseeable and fortuitous events by preventing others from exploiting their contingent situations of weakness.

In a study that constitutes the first true foundation of procedural justice theory, the objective was to determine what dispute-resolution system individuals preferred in the abstract. In the study, 84 subjects were given a menu of dispute-resolution systems describing their most salient procedural aspects but without making any direct reference to current legal terminology (*judgment*, *arbitration*, *mediation*, or *negotiation*), so as to prevent any cultural bias. The experiment was administered in such a way that there would be no confusion between judicial and nonjudicial procedures. The other significant aspect is that the subjects in one sample were placed under Rawlsian constraints: they were given no clue about what their role would be in the procedure (whether they were going to be the alleged perpetrator of a crime or its victim), and no clue about what evidence might be introduced against them or in their favor (Thibaut et al. 1974, p. 1275). This first group of respondents seemed to prefer an adjudicatory and adversarial procedure, in which the parties have the power to present their arguments and evidence and refute those of the counterparty, but where the final decision is taken authoritatively by a third and impartial subject.

To see whether preferences would change once the veil of ignorance was lifted, two further control groups were subjected to the same test: the subjects in one group were all assigned to the condition of victim of an unlawful act, are were given strong evidence in support of their case; the subjects in the other group were all assigned to the condition of alleged perpetrator of the illicit act, and were made aware of the possibility of an adverse decision. Under these conditions, it was found that preferences did change: the advantaged subjects in the victim group (the plaintiffs with strong evidence in hand) preferred an inquisitorial adjudication procedure (Thibaut et al. 1974, p. 1281).

This research indirectly supports the thesis that there exists an abstract sense of justice stipulating that in any given dispute it is fairer to confer a systematic procedural advantage on the disadvantaged party. Only under an adversary system where the principle of free party disposition is in effect would these disadvantaged parties be able to undertake a diligent search of the elements in their favor, trusting that these elements will have an impact on the final decision as a result of the greater control the litigants have on the procedure. In particular, the parties fear that the decision-maker will tend to rely on prejudice or will not take their personal position duly into account unless they can directly present evidence or arguments in their favor.

In subsequent research (Sheppard 1985), it was emphasized that, even though the option of an appeal against an unjust decision tilted preference toward an autocratic inquisitorial procedure, subjects generally preferred procedures over which they could exercise some degree of control. The objection that there may have been a cultural bias at work, making it difficult to generalize the results, was met by a subsequent repetition of the same study comparing its results with the results obtained by looking at French and German subjects. It emerged from this comparison that, just as individuals accustomed to a jurisdiction where the adversarial/disposition principle was prevalent were particularly averse toward the inquisitorial system, so individuals accustomed to an inquisitorial system were not particularly biased against the adversarial principle (Thibaut and Walker 1975, pp. 97–101, 120).

And yet, in studies comparing attitudes toward justice in non-Western societies it seems to emerge that there is indeed a difference in the way people perceive what is procedurally fair. Specifically, preference for adversarial adjudication procedures or for adjudication procedures based on the principle of party disposition is mainly a trait of individualistic rather than collectivistic societies (Gire 1997). In this regard, some differences have been found among those countries that consider themselves predominantly individualistic, and therefore place a higher value on autonomy, competition, outcomes, and self-sufficiency, as compared with those countries in which a collectivistic attitude prevails, and where a higher value is accordingly placed on harmony and group solidarity (Shelby Rosette et al. 2012).

In countries with an individualistic orientation, it is clear that the dominant values are better suited to adversarial procedures and the principle of party disposition, in that the individual wants to control the way the dispute unfolds and will not avoid competitive confrontation with the adversary. The difference emerges in comparison with Asian collectivist societies, where the adversarial system has been forcibly imported, but where in practice litigants have continued to largely defer to the judge for control over the proceedings. This has been observed in Japan, for instance, and the finding was subsequently confirmed in parallel observations of the ethnic Chinese population in Hong Kong, where litigation has been modelled on the strongly adversarial British system (Leung and Lind 1986, pp. 1135–1136): the ethnic Chinese population showed no specific preference for the adversarial elements in the procedure; if anything, it preferred a procedure that could guarantee the parties' ultimate control by excluding neutral third parties from exercising any formal decision-making power in the dispute, as in mediation. As much as Chinese sentiment did favor at least some degree of control, like the kind ensured by an adversarial procedure, they nonetheless did not view this aspect as critically important, or they felt that it was offset or outweighed by other procedural aspects deemed to be unpalatable, and strongly stigmatized in a collectivist culture, such as competition and direct confrontation with the counterparty (Leung and Lind 1986, pp. 1138–1139). On closer inspection, these results do undercut the criticism directed at the positions that culturalism and constructivism have supported as concerns the sense of justice. While the cognitive perspective has never denied the influences that culturally acquired preferences may have on the individual feeling of justice, it was by contrast in sociological and psychological constructivism that every biological-cognitive influence was excluded, claiming that "the social" wields full control over this area.

The foregoing studies on procedural justice have advanced our knowledge on the feeling of justice, giving us the first organic attempt to apply the methods of behavioral science to judicial procedures. The early attempts at studying procedural justice in this light are not, however, exempt from criticism. Damaška (1975), for instance, claimed that the research design was too simple, suggesting that the assumed models of judgment, arbitration, mediation, and negotiation did not in fact exist. However, this kind of research on procedural justice has provided fairly clear legislative-policy guidelines on the kinds of procedures that generally give some perception of control, as by making it possible for litigants to at least express their opinions. These findings about the preferences of parties in dispute make it possible to tailor judicial procedures accordingly (Shestowsky 2004).

This line of research on procedural justice found new impetus in the 1980s, clearly confirming that individuals appreciate control over the procedure, and this time the research was no longer solely based on laboratory simulations. Tyler and Lind's work contributes with a new remarkable insight: individual preferences formed *a priori*—that is, in the abstract and before we are actually involved in a proceeding—are not necessarily the same as when we are actively involved in a proceeding and have already experienced it. What is most striking is that in certain situations, our acceptance of the decision seems more closely linked to satisfaction with the perceived fairness of the procedure than to whether or not its outcome is favorable (Tyler and Lind 2001, p. 69). We would thus seem to be more interested in being treated with kindness and respect, and the outcome of the procedure would not be determinative in shaping our view of its fairness (Tyler 2002, pp. 23–34).

Hensler (2002, p. 93) asks why procedural justice is so relevant in judgments about justice. According to the *voice* hypothesis, subjects in a dispute prefer procedures that give them an opportunity to voice their needs and opinions. According to the *group value* or *relational* hypothesis, the disputants prefer procedural fairness because it gives them confirmation that they are in good standing within the social group, such fairness being equivalent to a sign of respect. Finally, according to the *fairness heuristic* hypothesis, because people fear being exploited by the group or by an autocratic decision-maker, but do not have the elements to understand whether this is happening, they rely on the following rule of judgment: if they perceive the procedure as fair, they have good reason to believe that nobody is taking advantage of them.

In a further phase, these studies have found that procedural fairness is not linked to particular venues of resolution or to particular disputes (Vidmar 1997, p. 129). Any layperson can distinguish between different types of procedures, and can appreciate that there is no procedure which can be universally the best in all situations. Five elements in particular stand out in a person's assessment of procedural fairness: feeling represented, ethical adequacy, interpersonal respect, neutrality, and the quality of the final outcome (Tyler 2002, pp. 21–22).

In other analyses (Berti 2002, p. 147), the idea of procedural justice is summarized in six principles: consistency in applying procedures, the decision-maker's ability not to be influenced, the accuracy of the information on which the procedure is based, the opportunity to review decisions, the degree to which the procedure represents the interests in question, and the ethical nature of the allocation of resources. In addition to trust in the institutions or confidence in the decision-maker's neutrality, the fundamental sentiment involved in procedural justice would therefore be our perception of our own social standing, that is, the position and respect we feel we enjoy within the group. The stronger the identification with a group, the greater the weight we attach to this particular aspect of being treated fairly (Berti 2002, p. 177).

Much of the research that looks at post-procedure individual satisfaction supports the idea of greater recourse to mediation (Alexander 2006; Ali 2018). Like any procedure in which the disputants exercise the greatest degree of control, mediation seems to satisfy various criteria of procedural justice. It would be for these reasons that mediation, once experienced, is more satisfactory than adversarial or inquisitorial procedures with a third-party decision-maker whose issuances are authoritatively binding, and this would justify even compulsory mediation (Berti 2002, p. 186), as is found today in several jurisdictions, where the parties are required to go through mediation before going to court (Schonewille and Schonewille 2014). Another study (Sheppard et al. 1988) sought to determine whether the litigants' preference may depend on the degree of conflict, testing the hypothesis that mediation seems fairer when the conflict is mild, but when the conflict intensifies the parties would favor arbitration-style procedures.

The policy-making rationale that emerged from the studies of procedural justice was that, for the welfare and satisfaction of citizens, judicial systems should provide fairer forms of participation and control over procedures, in addition to ensuring that the decision-maker is competent and the organization is efficient. As mentioned earlier, however, the paradigm of procedural justice has come under significant methodological criticism, which calls into question the validity of some concepts and so also the predictions made on the basis of the paradigm. For a more detailed account of these objections, we can first turn to the criticism that Taruffo has made of the validity of the survey methods used in determining the disputants' actual preferences. Taruffo has raised methodological doubts, especially as concerns the use of simulated situations and models, which did not make it possible to capture the complexity of real judicial experience (Taruffo 1981, pp. viii, xiii). He also argued that we cannot neglect the particular way in which the judge's role is perceived in each country's legal culture, speculating that the influence these conceptions exert is significant enough to make any kind of comparison fruitless (Taruffo 1981, pp. x, xvii).

The same line of criticism—but more detailed, and framed from the perspective of the economic analysis of law—is developed in the work of Kaplow and Shavell (2001), who argue that the choices expressed in laboratory simulations could not account for the intensity of these preferences; questions aimed at determining whether a given procedure was felt to be better than another could easily be understood in different ways, and depending on the respondent, they risked diverting attention to aspects other than personal preference and taste, such as one's immediate interest, or what constitutes good public policy.

The results of these studies on procedural justice also lent themselves to divergent interpretations, for they sometimes led to inconsistent results, as when subjects generally prefer direct participation but at the same time request someone (a lawyer) who will act as an intermediary between themselves and the decision-maker. Worse still, Kaplow and Shavell observed that the abstract procedural models used in the surveys did not properly reflect judicial and legal practice. On top of these criticisms, Hensler (2002, p. 92) argued that this body of research has been applied to fields that are too disparate (ranging from criminal law to civil and commercial disputes to decision-making procedures in organizations) or to situations in which there was no genuine conflict and no real demand for justice.

Worthy of mention among the substantialist objections, is Ricoeur's, quoted and endorsed by Sacchetti (2011, p. 340). According to Ricoeur, there cannot exist a purely procedural conception of justice, completely severed from the sense of justice that precedes it and accompanies it from beginning to end. In addition, part of the empirical research (van den Bos 1999; van den Bos et al. 1997) argued that the perception of justice can be subverted only to a certain extent by procedural aspects, and therefore satisfaction with the procedure itself will always be largely influenced by its distributive result. These positions underline the uselessness of a procedure which may be fair but is devoid of substantive justice, and which pursues "lowlier" goals that are more cosmetic than substantive (Berti 2002, p. 189).

On the assumption that the framework offered by the *distributive justice/procedural justice* dyad cannot cover everything, a third sense of *interactional justice* may be introduced in support of the first two. Interactional justice refers to what we perceive to be the fairness of the treatment we receive from others: it thus focuses on interpersonal sensitivity and respect rather than on the fairness of the decision-making procedure itself (Poon 2012, p. 1507). Interactional justice can in turn be broken down into a sense of "informational justice," relating to the quality and quantity of the information provided to explain the use of a given procedure, and "interpersonal justice," relating to respectful and dignified treatment (Beier et al. 2014, p. 47). But, as we have seen, the original notions of procedural justice already include this element, and so it seems to be that interactional justice can be a useful tool only in attempting to more accurately specify and analyze certain aspects of procedural justice that were previously treated as an aggregate. A similar point could be made in regard to retributive justice: although it represents a separate line of analysis, concerned in particular with criminal law and its sanctionative purpose, from an individual standpoint it also includes substantive allocation (in this case, of afflictive penalties), and it therefore delimits a distinctive aspect of distributive justice.

An even more convincing critique of procedural justice comes from the aforementioned study on its perception by defendants and the public (Beier et al. 2014). In the criminal proceedings taken into account, it was found that a judge's respectful behavior toward the accused had some impact on the public's perception of procedural justice, but not on that of the accused himself. Generally speaking, the perceptions of defendants were very distant from the public's perceptions, correlating with them in no more than a very marginal way. The researchers speculate that this may be a consequence of the audience's greater psychological distance from the accused, who would thus consider the situation from a less abstract point of view, ignoring the larger picture of procedural justice (van den Bos et al. 1997). This does not change the fact that if the research results were confirmed in other areas, the most relevant policy implications of the procedural paradigm would be undermined.

In conclusion, the difference between perceptions of procedural and distributive justice versus substantive justice would depend not on the nature of the perceived substance, but on the simple circumstance that the former would be able to emerge before the latter. In other words, our feeling of how we are treated in a procedure precedes our knowledge of its final outcome, and according to two principles that in cognitive psychology are referred to as the primacy and anchoring effects, this time differential would result in the temporally prior feeling (of procedural and distributive justice) being accorded greater weight in individual choice and evaluation (Arrow et al. 1995). In addition to being temporally pre-eminent, procedural correctness is also simpler to evaluate than substantive correctness. In fact, experiments show that the extent to which we appreciate the final result would actually influence our perception of the procedure's fairness, especially when the outcome was favorable despite the procedure being perceived as unfair (van den Bos et al. 1997).

The numerous positions and distinctions taken regarding our perception of procedural justice pose a significant problem in regard to what was becoming one of the cornerstones in the psychology of contemporary justice. Most of these, the ones just listed, are partly well founded, and the instruments of social research are probably still not suitable for analyzing such a complex package of perceptions, intervening variables, and feedbacks, like those involved in the continuous succession of formal steps in judicial procedures. The latter, in particular, are dynamic events, which in order to achieve a higher degree of precision would need to be broken up into different phases, entailing considerable difficulties in collecting and operationalizing the data.

However, many criticisms have failed to appreciate one aspect. Procedural justice is a specific area of study situated beyond an investigation of the factors that reinforce our perception of justice in the broadest sense: it has therefore never been intended to replace or absorb the fairness of distribution as an object of study. What is certain is that the need for respect, control, and social standing can greatly influence our perception of the fairness of a disputed allocation, in whatever way it is carried out. If research on procedural justice has not yet succeeded in demonstrating the phenomenological independence of procedural evaluations, it has fully demonstrated that they become part of the overall assessment of substantive justice. The acceptance of the idea of procedural justice leads to a significant consequence: our degree of spontaneous compliance with a normative system will always correlate as well with our satisfaction with perceived procedural justice in the process of gathering information and evidence, hearing parties and witnesses, and motivating judicial allocations and decisions (Tyler and Lind 2001, pp. 65, 71).

2.5 Transgression, Conformism, and Aggression

The concept of distributive justice is closely related to that of retributive justice. Our transgression and obedience of norms has been investigated philosophically for its free-will implications, but with the emergence of the behavioral sciences it has fully become an object of sociological and psychological investigation, the former focusing on exogenous factors, the latter on endogenous ones. The first modern study on the sociology of law was focused on anomie, with Durkheim's famous essay on suicide (1897). Psychology and sociology both attribute anomie to the absence of social control over instinctual drives. In psychology, the problem of observing norms is misleadingly examined in light of the simple influence of the social context and the social group, according to paradigms built on the analysis of social imitation and conformism (Conte 1997, p. xvi). In psychoanalysis, lack of conformity to what social structures require is thought to individually find its cause in our original nature, on the argument that conformity in essence grows out of a utilitarian calculus or out of nonrational conditioning. Only in a more mature phase have sociologists and economists been able to observe that some forms of opportunistic deviant behavior are actually psychologically "normal," just like conformist behavior

(Becker 1968; Merton 1968). Our perceptions of responsibility, severity, and merit, for instance, play a crucial role in determining our reactions to the penalties inflicted, as they become part of a chain of cognitive and affective responses (Berti 2002, p. 57). Even the idea of retribution, on which every wrongful act should have its appropriate punishment, is deeply rooted in the human soul (Donnerstein and Hatfield 1982, p. 317).

We have to go back now to the matter of conformism, understood as obedience to authority, an equivalence that does not quite stand up to scrutiny but remains central nonetheless. In her psycho-social analysis, Rosaria Conte (1997, p. 103ff.) distinguishes four kinds of obedience: (1) apparent, where conforming behavior derives from an independent volition, and only fortuitously coincides with what is prescribed; (2) instrumental, which includes various cases, such as imitation or an intention to appear diligent; (3) cooperative, where the agent subscribes to the aim pursued by the rule and thus follows the same rule; and (4) terminal, implying the activation of a meta-purpose or metanorm. Homans (1961, p. 293) asks, for example, what determines obedience to authority and identifies two fundamental factors: the nature of the order and the nature of the person who issues it. In regard to the latter, Homans took up a concept that was still relatively new in the social sciences: the concept of a charismatic leader, which he revisited in light of the modern concept of leadership, noting that the authority of leaders depends on their relationship with their followers, but also on these followers' relationship with one another. A leader's authority will thus depend, in no small part, on there being in the background a positive sense widely shared among followers, a sense that can be understood as a direct outgrowth of the established conditions of distributive justice. Feelings of justice correlate with resentment: when the latter is removed, the former tend to emerge (Homans 1961, p. 295). Authority, however, correlates with another element, namely, detachment from the authority figure's followers, with whom it is necessary to maintain a certain distance, because familiarity breeds contempt (ibid., p. 311). In stratified animal societies, however, the most significant episodes of rebellion come not from those at the bottom of the social pyramid, but from those who are near the top (Lopreato and Green 1990, p. 115), and in most cases this may also be true of humans.

The analysis of conformism, understood as an unconditional adherence to the majority, somewhat divorced from moral sense, is also relevant to understanding obedience to rules. In this meaning, conformism tends to be higher in collectivistic societies than in individualistic ones. In social psychology it is common to distinguish conformism from acquiescence. In the specialized language of psychology, acquiescence is a behavioral response to another individual's request, while conformism refers to the group's influence on an individual. The methods of acquiescence are ingratiation, appeal to reciprocity, or leverage on the sense of guilt (Hogg and Vaughan 2011, p. 230). Those we call conformists tend to have low self-esteem, a strong need for support or social approval and for self-control, a lower IQ, a sense of anxiety and guilt, feelings of inferiority and insecurity within the group, and an authoritarian personality (ibid., p. 250). In this regard, it is necessary to cite the study by Stanley Milgram (1963), who in the general population experimentally found a surprisingly high degree of obedience in the form of conformism/acquiescence. These now-paradigmatic experiments showed that almost two-thirds (65%) of a sample of middle-class adult males did not have much hesitation administering 450-volt electric charges to another individual, this simply acting on the orders of an experimenter who did not have any real coercive power (the subjects were administering these charges without knowing that they were not in fact real but only simulated).

By accepting to take part in the experiment, the subjects implicitly accepted to inflict this kind of suffering on other individuals for reasons that were little more than futile, namely, to reward or punish them for correctly or incorrectly answering certain test questions. And even a few months after the experiment, the vast majority of the subjects who had administered the electric shocks as punishment on precisely that premise did not express remorse for what they had done. It was also found in the experiment that obedience to the orders diminished (a) the more the victim's suffering increased (with the victim simulating greater pain), (b) the more it became apparent that the victim could actually see the action of administering the charge, and (c) the more the subject felt responsible, as a result of grasping the relation between his actions and their consequences.

It was thus found that the subjects' willingness to break with conformism and rebel against submission derived not only from seeing the suffering they were inflicting on the victim but also from the perception they would form of themselves as a result of being visible to their victims. In this phase, therefore, it was the individual's functioning in the social situation that redefined the subjects' cognitive field and their relationship with the victim (Ravenna 2012, p. 150). Milgram's research also showed that obedience diminished as the authority's status diminished, and so was a response to authority itself rather than its orders. Disobedient subjects tended to consider themselves more responsible for the sufferings inflicted, while obedient ones tended shift responsibility onto the victim. Another instance of conformism was found through Asch's experiments in social psychology. When asked to offer a personal judgment, the subjects easily changed their opinion (about half the time) if confronted with a majority of peers or with a group of experts opposed to their opinion (Asch 1952). The clear psychological propensity to imitate the behavior of the closest social group was also highlighted by Hodgson and Knudsen (2010, p. 160).

Our spontaneous tendency to obey, and the education we receive in obedience, brings on a general tendency in the population to feel responsible for obeying the orders of authorities, but not for the consequences that such obedience carries. This could be defined as the *conformist scope* of the authority-obedience relationship. Our relationship with someone giving orders from a position of authority (in Milgram's experiment authority is personified by the researcher who gives orders and oversees the experiment) seems to lend credence to Hannah Arendt's insight about the "banality of evil." In a bureaucratic and depersonalized context in which our social role is to follow the directives issued by a power that exercises control and oversight, we may perceive our actions to be neutral and routine even when we would otherwise consider them inappropriate and even immoral. It could also be objected that an authority's order does not necessarily follow the rules or the spirit of a legal system's most fundamental and binding norms, and that it only partly coincides with the feeling of justice and with our cognition of the law. Each non-overly conformist subject comes intuitively to postulate the existence of an escape clause that makes it permissible, and sometimes even obligatory, to disregard orders that are illegitimate, either because they are contrary to some core right or because the acts they impose are felt to be repugnant.

Milgram goes through the mechanisms that bring the internal conflict between consciousness and conformism to a tolerable level. These are the behavioral subterfuges he observed in his experiments on the "conformist routine": he found that his subjects would (i) isolate themselves from unpleasant events by looking the other way, and reading the instructions aloud to cover the screams and focus on the rules of the experiment; (ii) ignore the evidence so as to adopt a more comforting interpretation that denies the victim's pain or the strength of the electric charges; (iii) carry out orders only in part, by administering less painful shocks and showing benevolence toward the victim, or by devising stratagems designed to suggest the right answer to the victim (thereby sabotaging the experiment, while avoiding a frontal defiance of authority); (iv) reduce their sense of personal responsibility by asking the experimenter for reassurance that they would not be held responsible, and putting the blame on the victim; (v) release their tension by way of psychosomatic reflexes (sweating, anxiety, trembling, hysterical laughter); and (vi) express dissent or disapproval of the action ordered by the experimenter, all the while continuing to perform the assigned task.

Other research carried out subsequently found that in obedience, control over our actions and our responsibility for them is decentralized to a source that resides outside us and is pinned to those actions themselves (Ravenna 2012, p. 166). Aggressors often use their relationship with the victim, and the victim's personal characteristics, as a lever by which to claim that they are not to blame. In fact, it is common for the aggressor to belittle the victim. As victims of an assault or an injury, we may unknowingly distort reality to justify our own victimization: this is less traumatic than admitting that the world is fundamentally unjust and that we are powerless to do anything about it. As victims of an aggression, we tend to console ourselves by imagining that we have not suffered any injury or that the injury will be indemnified by other compensatory advantages (Donnerstein and Hatfield 1982, p. 320).

In tracing out the origins of disobedience, Conte instead describes the social dynamics of rule transgression. Particularly comprehensive and illuminating is her analytical examination of "social anomie" (Conte 1997, pp. 194–199). Disobedience can spread by "positive reinforcement," that is, as a result of impunity for another transgression. A similar mechanism is involved when disobedience spreads "by generalization" and "conformity," as happens when our observance of an implicit norm makes compliance with an explicit rule unnecessary, as in the example of jaywalking, in which we break the explicit rule against crossing the street with a red light or outside the crosswalk, on the reasoning that we are following an implicit norm under which this behavior is legitimate if widespread and if safe under the circumstances.

From these forms, we move on to what Conte calls the "analogical or innovative" spread of transgression—where transgression is favored by contiguity among the areas where violations occur—and then on to its spread by "forced compliance," where the transgression proves necessary to avoid the greater risks posed by obedience. Consider, for example, the social costs we incur, especially the toll taken on time, by complying with the myriad bureaucratic rules which are put in place to ensure that our transactions with government agencies are open and transparent, but which in time grow outdated, rendered useless by technological innovation, or which only wind up providing these agencies with a constant stream of information and data they already possess.

Consider also the excessively prudent road-safety regulations that only wind up making the road *less* safe, as in the example of the speed limits which government agencies in some countries put in place to avoid the risk of legal action, but which are unreasonably low relative to the braking, impact-protection, and road-handling capacities of the vehicles currently in circulation: often there is no safety benefit that comes with these rules, which in fact may lead to accidents that would not happen but for the compliance itself. Finally, disobedience may spread through the "run-up effect" (as when people in the stands of a stadium get up, forcing those sitting behind them to also stand up) or by "lack of social control," when victims have colluded or consented, or when they are thought to have deserved what they got.

The dynamics described by Conte are not mutually exclusive, and in some cases they raise doubts because they operate on different planes of analysis, at times appearing to confuse the root causes of disobedience with their mediating variables, or even simply with their background or reinforcement factors. In the field of taxation, for example, Ferrari and Randisi (2011, pp. 142–143) use a large body of empirical research to argue that the effect a tax obligation has on our perception of justice, and on subsequent decisions about whether to comply, is mediated on an individual level by our *affect intensity*, that is, by our predisposition to be strongly influenced by our own emotional states and therefore to perceive anger to a greater extent. Conte does, however, continue her analysis by highlighting a fundamental aspect in legal attitudes, namely, the adaptation mechanism through which the transgression of a norm becomes bearable or tolerated in the long run. Conte (1997, pp. 199–201) hypothesizes that very often a precise social-cognitive mechanism is at work on which basis transgression engenders further regulation, which in turn engenders further transgression.

Adaptation mechanisms may be found in different forms: adaptation by way of artefacts, such as anti-theft devices and other security systems whose generalized use for certain purposes makes it acceptable for a rule to be broken when those who are harmed should have taken those protective measures but failed to; adaptation by way of rules of conduct designed to do the job of adaptation by way of artifacts, here too supporting the view that any harm you suffer as a result of failing to behave according to the rule makes you a careless victim and hence an aider and abettor of sorts; adaptation by way of conventions, which consists in the practice of attenuating the gravity of violations by taking precautions meant to make those violations socially more acceptable, such as parking on the sidewalk but taking care to leave enough room for pedestrians to walk through.

Some psycho-social mechanisms operate specifically when transgression and disobedience prompt an aggressive reaction, as when someone's sphere of inviolability is frontally and explicitly breached. The hypotheses formulated in the human sciences in regard to the reasons for aggression do not necessarily illuminate the path of law. More convincing, perhaps, is the attempt of ethologists and evolutionary biologists, who see aggression as a hereditary trait that in many situations affords an adaptive advantage to those who carry it, because it gives them the ability to react to specific adverse stimuli posed by the environment. It would be our very nature as evolved primates, then, that would favor aggression, as shown by the modern refutations of the myth of the noble savage, with the discovery that the majority of anthropologically intact populations engage in some form of endemic and widespread conflict (Cunliffe 2006, pp. 67–68). Human beings practice extremely sophisticated forms of social aggregation, but at the same time have innate competitive and aggressive traits.

On closer inspection, however, aggressiveness proves to be different from aggression. An actively or potentially threatening behavior does not necessarily consist in violating the protected sphere of others, thereby breaking the law. The neurological difference between reactive aggression (an uncontrollable and often misdirected violent impulse) and instrumental aggression (exemplified in its rational extreme by premeditation) reflects, for example, the different directions that decisions take in cognitive-instinctive systems rather than in cognitive-reflexive ones (Blair 2007, pp. 322–323). Reactive aggression is often triggered by frustration or a threat, and is unleashed without the subject having a specific purpose in mind. Instrumental or proactive aggression is engaged in to achieve a specific objective, which can be to appropriate something or to gain status in a group's hierarchy.

Aggressive acts are caused, for example, by physiological stimuli such as swelter or overcrowding. Individual "frustration" (Hogg and Vaughan 2011, p. 459ff.) has been hypothesized as the main cause of aggression, but the objection seems valid to me that not all frustrating situations give rise to aggression, while the force inhibiting an act of aggression correlates positively with the level of anticipated punishment as a consequence of that act or with its expected social disapproval (Donnerstein and Hatfield 1982, p. 322). The theory of social learning suggests, on the contrary, that acts of aggression are primarily a result of the socialization processes that take place in childhood. Individuals are in this sense thought to be aggressive because they learn to be that way as children. Through imitation, they learn to attack like adults, and they are then directly or indirectly rewarded for their aggressive actions (Hogg and Vaughan 2011, p. 462).

The concepts of primary and secondary socialization have been critically assessed and reviewed from a perspective according to which our attitude toward rules and laws develops mainly through our interactions (Favretto 2016, p. 211ff.). Mixed models have also been proposed that combine elements of bio-psychological and social activation. An example is the excitation-transfer model. This model predicts that a subject will escalate to acts of aggression when any of the following three factors recur: a learned aggressive behavior, an element of activation or excitement from any source, or an interpretation that evaluates aggression as an appropriate response (ibid., p. 460ff.).

Modern ethology has shrewdly changed the focus of aggressive action from an internal state to a product of a conflict of interest (de Waal 2000). Aggression is the outcome of a social decision, and it represents a particular way of resolving conflicts between groups or individuals. On the basis of this "relational" hypothesis, we can make predictions about the occurrence of aggressions, which should increase along with an increase in opportunities for competition, in the value of the resources at stake, or in the likelihood that the aggressor's relationship with the person aggressed against can be recovered, while it should decrease with any increase in the risk of injury or in the value of the relationship (de Waal 2000, p. 27). De Waal also notes that, paradoxically, it is precisely when resolution mechanisms work best that individuals will be less reluctant to engage in open conflict.

On the other hand, aggression should not be considered an inherently negative force, and provided that some mechanisms for controlling and mitigating the repercussions are in place, it could well be a component of social relationships (Aureli and de Waal 2000, p. 5). Pinker (2011, p. xiv) argues that violence, all things considered, has been declining over the course of human history, and that aggression should not be viewed as a single, episodic urge but rather as "the output of several psychological systems that differ in their environmental triggers, their internal logic, their neurobiological basis, and their social distribution."

2.6 The Criticisms Against Legal Psychology, and Why They Should Be Rejected

As has been noted in social psychology, the "cognitive emergence" of legal norms has made obsolete the attempt to demonstrate their original usefulness (Conte 1997, p. 80). The individual's obedience can be explained by reference to the "artificial" utility of law, which promotes compliance without an understanding of the original reasons for a rule or of its actual usefulness. In dealing with a legal system, whatever its form, the individual is instinctively led to think, "If this is the rule, there must be a reason for it," and even to actively investigate what this reason might be. We have seen that this is often a source of tragedy and absolutism, when obedience morphs into uncritical conformity, but there is no doubt that conformity to established norms, or at least the appearance of conformity, brings social advantages that cannot be discounted.

If at this point it seems beyond doubt that psycho-social factors play a role in shaping our idea of justice and individual legal attitudes, it is necessary to organically respond to the main criticisms directed against the very possibility of founding a "psychology of justice" in the proper sense, and to the thesis that the sense of justice is at bottom social and can therefore be fully explained in social terms alone. Edward Sampson (1983) argues that what is being billed today as a psychological understanding of justice is in reality the reflection of a social status that instead shows a *lack* of justice. The first point of attack is Lerner's (1980) "just world" hypothesis, according to which individuals develop the belief that there is a more or less ineluctable correspondence between what they have and what they deserve. In essence, we tend to be convinced, often without acknowledging it, that people ultimately deserve what happens to them, because there is a higher criterion of justice at work.

This "just world" belief would be rooted in a basic psychological need to see the world as a just place. Lerner does not clearly place this mechanism in the biological or in the social field. According to an explanation of the first type, we should have to hypothesize the presence of a "just world" gene or set of genes that would provide their carriers with an unconscious self-defense and coping mechanism by which to deal with the existent. This hereditary trait would in many situations enable us to endure adversity by inducing us to construct such adversity as the result of a higher justice beyond our ken. This would also exclude the idea that our misfortune should derive simply from chance, and would predispose us to hope that the situation will improve if we can behave differently. It cannot escape us how a mechanism so described is the cornerstone of almost every system of religious beliefs. According to an explanation of the second type, we would have to hypothesize that our fundamental belief in a just world is simply the worldview the ruling group proffers as a way to legitimize its own power. Here, too, we have a system of quasi-religious beliefs, a scheme through which the dominant group legitimizes its dominance in the eyes of the dominated group. This group would internalize the scheme and thus come to accept its own position as substantially just and deserved.

Sampson observes that this belief can be dysfunctional from the standpoint of both distributive and procedural justice. If we were truly in the grip of the just-world belief, we should have to embrace fatalism or acquiesce in any kind of substantive injustice. Sampson, however, reads Lerner's theory through a strictly sociological lens, arguing that belief in a just world is socially constructed to legitimize social injustice (Sampson 1983, p. 3), and every paradigm based on biological adaptation is accordingly analyzed by him as instrumental psychologization. To make his case, Sampson uses the example of our perception of justice in the workplace, where it would be clear that not only our evaluation of personal investments and outcomes but also our reactions to injustice are psychologized and reduced to the individual dimension (ibid., p. 12). This example can be used to argue that the use of the individual and interpersonal dimension as a point of departure and arrival carries the risk of distorting the global understanding of the phenomenon, and of losing sight of "the largest socio-historical totality" (ibid., p. 17).

By taking an openly critical perspective, Sampson relegates the psychological component to a superstructural role. Only through an understanding of history and society can we come to an understanding of the feeling of justice, while the purely psychological perspective implicitly serves potentially ideological purposes (ibid.,

p. 19). In particular, our conceptions of equity and of equality would both be historically determined by their context, and could not be analyzed as timeless psychological principles of human behavior (ibid., p. 36). This fundamental error is attributed by Sampson to the individualistic bias of Western societies and to the myth of the *self-contained person*. It should be clear that this is not simply an *expression* of natural individualism but an *extremization* of it (ibid., p. 47).

As concerns the idea of justice, Sampson follows the same steps as those taken in forming the constructionist hypothesis in social psychology (Gergen 1985). The psychological study of the processes by which the idea of justice is constructed would mainly consist in an idiographic exercise in historical analysis and in a psychological analysis of norms that are are not "basic" in the sense of the natural sciences but rather cultural (Sampson 1983, p. 57). According to Sampson's idea, the individual subject is not reducible to a material fact, it is a historical, social and cultural creation, that is, a mediated object, or the result of a complex of underlying social structures and principles that define what it means to be a person. Hence the denunciation of the cognitivist vision which has become predominant in psychology. Cognitivism views internal mental processes as a fundamental lens through which to understand human behavior, ignoring its social processes of production (ibid., p. 87).

Sampson (1983, pp. 111–113) takes some paradigms that in his view misconceive the relation between psychology and society, and distills them down to three classes, identifying an "impotent," a "purist," and a "liberal" position. Among these the purist position is singled out in particular as a case of obtuse positivism. This criticism bears strong similarities to that which holistic macrosociologies had levelled at sociological paradigms (extensively discussed in Chap. 1) which look to individualism to support the possibility of a real distinction between facts and values. But the same criticism is also directed at the liberal position, for even if this position recognizes that values end up influencing the choices we make in selecting an object of study, it would pretend to objectivity in ascertaining facts and testing its hypotheses: as Sampson argues, if our prejudices are unconscious and ideological, and therefore inaccessible to us and our contemporaries because they are socially determined, we cannot expect to get rid of them in the validation phase.

Sampson goes on to claim that the psychology of contemporary justice has fallen victim to what he calls "truncated subjectivism," meaning that it tends to define justice from the personalistic perspective of individual phenomenology. This individualism loses sight of the overall functioning of justice within society, and once more proves to be ideological and limited. The charge of abstract individualism is also levelled at *psychologicization* (it is not clear why this term invariably takes on negative connotations): it attaches to the emphasis on individual subjectivity, which Sampson equates with a negation of intersubjectivity. In Sampson's view, these tendencies lapse into an uncritical empiricism—which confines itself to description, without considering what changes might have to be made—and into post-allocation analysis, in which the distribution processes that have already come to pass are taken as a foregone conclusion (Sampson 1983, pp. 147–148). If the conditions of justice, such as they exist in society, are uncritically accepted, no less uncritical will be our analysis. This analysis will perpetuate those same concepts, rooted in the idea of a market economy, so as to lend social legitimacy to standards of synallagmatic exchange and proportionality (ibid., pp. 150–153).

Nor is Sampson entirely convinced by the *address-frame* approach, which mediates between social and biological elements, and in this way situates the analysis of the sense of justice within social interaction. According to this paradigm, justice refers to the way in which our actions become intelligible: these are actions that we and others can evaluate without referring to any state of internal need (ibid., p. 174). It is the very context of actions that defines the relative sense of justice. By introducing contextual factors, the subjects involved in a relation of justice develop a course of action in light of the knowledge held by those with whom they interact. In this way we would gain a clearer understanding of altruism, the sense of justice, dissonance, attribution, and aggression: these are not so much attributes of the psyche as interpersonal processes driven by a motive. People are interested in their own self-presentation and self-image for two reasons: to gain some reward and to support an ideal concept of themselves. The characterization of justice as an address-frame also makes justice negotiable (ibid., p. 178). The range of socially legitimate conceptions of justice is the range of those that can be legitimately negotiated. Yet even here we cannot succeed in the attempt to implant a "natural" motif in the individual interest in self-esteem, hedonistic satisfaction, or control over the external environment, implying a basic instinct or a cognitive interest that is part of human nature.

Naturally linked to this criticism is the objection to the idea of procedural justice (ibid., pp. 184–185): as discussed, the objection is that this idea separates the procedure from its outcome, when justice should instead establish their unity in the context of the social relationship. It seems to me that the criticisms aimed at positivism, and perhaps at the very possibility of constructing an objective science of society, have been sufficiently addressed in Chap. 1. Even so, it is worth underscoring that there is a fallacy in Sampson's culturalist position, for he deprives scientific investigation of the possibility of describing and predicting reality in universally comprehensible terms. If this were the case, we would not be able to explain why those parts of ancient or medieval science based on assumptions and beliefs that would later turn out to be wrong still remain perfectly understandable to us, even if they arose in a radically different social, economic, and cultural context. In psychology, too, the incomparability of scientific paradigms that originated in different social contexts proves to be an ideological cover for a militant science. Psychology has certainly for some time ignored the importance of culture in its attempt to understand human behavior, recognizing culture as serving no more than a superficial function in explaining the functioning of the individual psyche (Miller 2001, p. 23), but this lack of interest seems to have been overcome. This development can be appreciated in the evolution of sampling techniques, which are now more refined and can face a multiethnic and culturally diverse research universe, as well as in the theoretical effort to integrate the results of cultural anthropology with psychological insights. In parallel, and in a coherent manner, a greater emphasis has been placed on evolutionary theory and other biological approaches in an effort to broaden the individualistic paradigm (ibid., p. 22). The evolutionary principle seems to have definitively established itself in the anthropological debate (Dunbar 2009; Nettle 2009; Runciman 2009). I will later come back to the *cognitive turn* that can now be observed in all fields of psychology and social psychology, including those research fields that deal with justice, and several of these theories will be explored in Chap. 3.

If we return now to the objections that Sampson raises against the psychology of justice, we can reasonably make this assessment in light of the previous discussion, namely, that the charge of the ideologism implicit in the empirical-positivist paradigm ultimately comes back to bite those who formulate it. Sampson moves from a line of critical thinking that hides a fallacy, because it invents concealed and unfalsifiable virtual agencies, and even more importantly because it ignores the possibility that social action should give rise to unintended consequences (Conte 1997, p. 204). I will later provide a stronger argument for the view that a psychology of law properly so called can serve an important function in the analysis of legal attitudes (in both its psycho-sociological inflection and, more clearly, its bio-psychological one). But it seems useful to first define some additional concepts that are essential in deciphering the structural foundations of psychological constructionism.

The gist of this latter position, most paradigmatically associated with Gergen (1985), is that theoretical propositions that contain mental predicates—that is, propositions that are entailed by the existence of a link between mental predicates and reality—can neither be true nor false. Even the non-psychological statement that the earth is spherical and not flat would be neither true nor false in terms of its figurative value, that is, in terms of its correspondence to the real world. According to the social constructionist model, the validity of a psychological theory cannot be assessed by its correspondence to empirical facts. In this critique of the positivist approach to psychology, the psychological aspect is reconceptualized as social, and the social is defined as a historic unity. Facts, knowledge, and psychological theories are socially constructed and constituted through social processes and conventions. The traditional categories of cognition, motivation, emotion, and learning would not be real properties in the individual psyche but would rather be set in discourse, and this would be central to the constitution of social reality. As a result, social psychology can only provide historicized answers, that is, answers linked to their context and to socio-linguistic conventions.

It is also worth looking at the objection that socio-psychological constructionism has received from Hibberd (2005). Her stance is in my view essential to a cognitive sociology of law, because it reveals that this form of constructivism is very different from sociological creationism. According to Hibberd, social constructionism in psychology replicates the problems of sociological positivism itself by forging a metatheory based on different disciplines such as the philosophy of language, the sociology of knowledge, literary analysis, hermeneutics, social phenomenology, and anthropology. Social constructionism suffers from an underlying inconsistency because it uses empirical evidence to support its conclusions, which would then paradoxically delegitimize any empirical and positivist approach (Hibberd 2005, p. 2).

Despite these challenges, however, the appeal of constructionism is remarkable, in sociology as in psychology, because it is almost limitless and boundless in the perspectives it makes possible. An idea we commonly find in constructionist conceptions is that scientific work need not yield verifiable theories (ones whose propositions must be found to be either true or false) but should rather provide us with a toolbox of linguistic resources that may be assessed simply as instructive or not, as credible guides for scientific inquiry, as stories that "work," rather than as true or false accounts (Hibberd 2005, p. 11). Quite correctly, Hibberd describes this as a "semantic holism" that postulates the incommensurability of theories. It proceeds from a rejection of meaning as having an external referent, and from a view of it as entirely indeterminate, contextual, and social (ibid., p. 16). Hibberd's interesting thesis, in social psychology as in sociology, is that constructivism is not actually antithetical to positivism in the social sciences. On the contrary, its underlying trend is parallel and assimilable to the latter. In the empirical psychology that originated out of positivism, attention gradually but inevitably shifted toward nonsubstantial details, leaving out the general theoretical implications and ignoring the conceptual contradictions that sooner or later would have become overwhelming. The same thing happened to psychology of a more theoretical and philosophical bent (ibid., p. 180).

Ehrenzweig reminds us of the ill-conceived mistrust that philosophers have of psychology (which, almost ineluctably, includes the mistrust that legal philosophers have of the psychology of justice), an aversion shared in political science but also in the anthropology and sociology of law. By refusing to deal in psychological investigation, these disciplines have protected their field from semantic and emotional games, including through their choice to take only what is conscious and verifiable into account as worthy of analysis (Ehrenzweig 1971, pp. 158–160). Pattaro (2005, p. 390) likewise highlights the negative effects of the reluctant attitude that Anglo-American analytic philosophy has traditionally had when it comes to drawing on the empirical findings of psychology.

The psychology of law has so far confined itself, in academia, to criminal law and criminology, where it was certainly easier to understand the way personological and psychological factors bear on legal decision-making, and where the problems of judicial efficiency became immediately apparent when investigating crimes and profiling criminals (Justickis 2008, p. 234). In fact, as a result of misdirected specialization, law has been loath to open its doors to the other social sciences, and certainly it has not appreciated their advances, thus remaining wedded to sociological or psychological paradigms that are now largely discredited; or, at best, it has been selective and self-serving seeking and using their insights. Jurisprudence has often explicitly objected to the empiricism of the psychological approach, criticizing the methodology of these studies or denying the possibility of even partly basing a judicial decision on these results (Tanford 1990, pp. 145–148).

The psychology of law has, for the most part, remained the province of researchers with strictly psychological training, when the most important need was to tear down barriers (McEwan 2006, p. 29). I agree with the proposition that law cannot be based on reasoning alone, nor can it be reduced to emotionality alone (Goodenough

2006, p. 90). To be sure, there are cases in which it is the legal norm itself that uses intuitive processes by basing judgments on specific patterns of behavior or attitudes that cannot be typified under any general rule (modesty, honor, diligence, best effort). But this should not be taken to mean that law and psychology can be considered two sides of the same hypothetical "science of the mind," such that the more implicit and naive the former (law), the more explicit the latter (psychology) (ibid., p. 77).

As in Chap. 1, I will instead support the need for an integrated approach to these problems by using different methods of analysis, an approach I once more will qualify as cognitive. The cognitive approach has pragmatically shown that we should neither accord primacy to the social or the cultural over the biological nor to do the opposite. The emphatic expression "cognitive revolution" (Miller 2001, p. 27) simply points out that mental processes are also cultural processes. The research done on patterns, biases, and any other cognitive structure has enabled us to appreciate how we influence our affective and behavioral reactions by attributing meaning to our own experiences.

It is precisely in cognitive neuroscience that the foundational role of psychology in the study of law has been highlighted, and where we can raise questions that seem so pertinent to any analysis of juridical behavior and to cognition of the law (Goodenough 2006, p. 79): what happens in our brains and minds when we frame rules, when we take these rules into account in making decisions about what to do or about our past actions, and when do we apply these rules in judging the behavior of others? Legal obligations need to be situated in the mind. Otherwise we will not be able to fully understand the origins and effects of rule-based behavior (Conte 1997, pp. xvii–xix) or formulate an adequate "theory of normative cognition" (ibid., p. xxii) that can above all explain behaviors of spontaneous compliance. Such compliance, while not driven by any distinctly moral sense, is nonetheless authentically normative. One answer to this question may lie in the "cognitive emergency" of norms. This idea will be developed in Chap. 3, where we will look at several sociobiological hypotheses by which to explain legal attitudes from an evolutionary-cultural perspective.

References

Adorno, Theodor Ludwig, Else Frenkel-Brunswik, Daniel Levinson, and Sanford Nevitt. 1950. *The Authoritarian Personality*. New York: Harper & Brothers.

Alexander, Nadja. 2006. Global Trends in Mediation: Riding the Third Wave. In *Global Trends in Mediation*, ed. Nadja Alexander, 2nd ed. Alpheen an den Rijn: Kluwer Law International.

Ali, Shahla F. 2018. *Court Mediation Reform: Efficiency, Confidence and Perceptions of Justice*. Cheltenham: Edward Elgar Publishing.

Allen, Vernon L. 1982. Effects of Conformity Pressure on Justice Behavior. In *Equity and Justice in Social Behavior*, ed. Jerald Greenberg and Ronald L. Cohen. New Haven and London: Academic Press.

Amsterdam, Anthony G., and Jerome Bruner. 2000. *Minding the Law*. Cambridge, MA, London: Harvard University Press.

Ariely, Dan. 2009. *Predictably Irrational*. New York: HarperCollins.

Arrow, Kenneth, Robert H. Mnookin, Lee Ross, Amos Tversky, and Robert Wilson, eds. 1995. *Barriers to Conflict Resolution*. New York, London: W.W. Norton & Company.

Asch, Solomon E. 1952. *Social Psychology*. Englewood Cliffs, NJ: Prentice-Hall.

Aureli, Filippo, and Frans B.M. de Waal, eds. 2000. *Natural Conflict Resolution*. Berkeley, Los Angeles: University of California Press.

Banaji, Mahzarin R., Kristi M. Lemm, and Siri J. Carpenter. 2001. The Social Unconscious. In *Blackwell Handbook of Social Psychology: Intraindividual Processes*, ed. Abraham Tesser and Norbert Schwarz. Malden, MA: Blackwell Publishers Inc.

Bandura, Albert. 1986. *Social Foundations of Thought and Action: A Social Cognitive Theory*. Englewood Cliffs: Prentice-Hall.

Becker, Gary S. 1968. Crime and Punishment: An Economic Approach. *Journal of Political Economy* 76 (2): 169–217.

Becker, Howard S. 1963. *Outsiders: Studies in the Sociology of Deviance*. New York: Free Press of Glencoe.

Beier, Susanne, Constanze Eib, Verena Oehmann, Peter Fiedler, and Klaus Fiedler. 2014. Influence of Judges' Behaviors on Perceived Procedural Justice. *Journal of Applied Social Psychology* 44 (1): 47.

Berti, Chiara. 2002. *Psicologia sociale della giustizia*. Bologna: Il Mulino.

Biernat, Monica, and Laura S. Billings. 2001. Standards, Expectancies, and Social Comparison. In *Blackwell Handbook of Social Psychology: Intraindividual Processes*, ed. Abraham Tesser and Norbert Schwarz. Malden, MA: Blackwell Publishers Inc.

Blair, R.J.R. 2007. Aggression, Psychopathy and Free Will from a Cognitive Neuroscience Perspective. *Behavioral Sciences and the Law* 25: 322–323.

Bohner, Gerd, and Norbert Schwarz. 2001. Attitudes, Persuasion, and Behavior. In *Blackwell Handbook of Social Psychology: Intraindividual Processes*, ed. Abraham Tesser and Norbert Schwarz. Malden, MA: Blackwell Publishers Inc.

Bona, Carlo, and Rino Rumiati. 2013. *Psicologia cognitiva per il diritto: Ricordare, pensare, decidere nell'esperienza forense*. Bologna: Il Mulino.

Bosisio, Roberta, and Roberto Cammarata. 2007. Conclusioni. In *Non è giusto! Dilemmi morali e senso di giustizia nelle rappresentazioni degli adolescenti*, ed. Paola Ronfani. Rome: Donzelli.

Boudon, Raymond. 2009. *Effets pervers et ordre sociale*. Paris: Quadrige - PUF.

Bowers, William J., and Glenn L. Pierce. 1980. Deterrence or Brutalization. *Crime & Delinquency* 26 (4): 453–484.

Brooks-Gordon, Belinda, and Michael Freeman. 2006. Law and Psychology: Issues for Today. In *Law and Psychology*, ed. Belinda Brooks-Gordon and Michael Freeman. Oxford: Oxford University Press.

Burnstein, Eugene, and Christine Branigan. 2001. Evolutionary Analyses in Social Psychology. In *Blackwell Handbook of Social Psychology: Intraindividual Processes*, ed. Abraham Tesser and Norbert Schwarz. Malden, MA: Blackwell Publishers Inc.

Canter, David. 2008. In the Kingdom of the Blind. In *Psychology and Law: Bridging the Gap*, ed. David Canter and Rita Žukauskienė. Aldershot-Burlington: Ashgate.

Canter, David, and Rita Žukauskienė, eds. 2008. *Psychology and Law: Bridging the Gap*. Aldershot-Burlington: Ashgate.

Cohen, Ronald L., and Jerald Greenberg. 1982. The Justice Concept in Social Psychology. In *Equity and Justice in Social Behavior*, ed. Jerald Greenberg and Ronald L. Cohen. New York, London: Academic Press.

Conte, Rosaria. 1997. *L'obbedienza intelligente*. Rome, Bari: Laterza.

Cook, Karen, and Karen Hegtvedt. 1992. Empirical Evidence of the Sense of Justice. In *The Sense of Justice: Biological Foundations of Law*, ed. Roger D. Masters and Margaret Gruter. Newbury Park, London, New Delhi: Sage.

Crespi, Franco. 1989. *Azione sociale e potere*. Bologna: Il Mulino.
Crosby, Faye, and A. Miren Gonzalez-Intal. 1982. Relative Deprivation and Equity Theories. Felt Injustice and the Undeserved Benefits of Others. In *The Sense of Injustice: Social Psychological Perspectives*, ed. Robert Folger. New York, London: Plenum Press.
Cunliffe, Barry. 2006. The Roots of Warfare. In *Conflict*, ed. Martin Jones and A.C. Fabian. Cambridge: Cambridge University Press.
Daly, Martin, and Margo Wilson. 1997. Evolutionary Social Psychology and Family Homicide. In *The Maladapted Mind. Classical Readings in Evolutionary Psychopathology*, ed. Simon Baron-Cohen. Hove, New York: Psychology Press.
Damaška, Mirjan R. 1975. Presentation of Evidence and Fact Finding Precision. *University of Pennsylvania Law Review* 123: 1083–1106.
Dauer, Edward A. 2006. Strong Reciprocity and Accountability: Behavioural Analysis of Patients' Legal Responses to Medical Injury. *International Journal of Law in Context* 2 (3): 257–276.
de Waal, Frans B.M. 2000. The First Kiss. Foundations of Conflict Resolution Research in Animals. In *Natural Conflict Resolution*, ed. Filippo Aureli and Frans B.M. de Waal. Berkeley, Los Angeles: University of California Press.
Deutsch, Morton. 1985. *Distributive Justice: A Social-Psychological Perspective*. New Haven and London: Yale University Press.
Di Donato, Flora. 2008. *La costruzione giudiziaria del fatto: Il ruolo della narrazione nel "processo"*. Milan: FrancoAngeli.
Donnerstein, Edward, and Elaine Hatfield. 1982. Aggression and Inequity. In *Equity and Justice in Social Behavior*, ed. Jerald Greenberg and Ronald L. Cohen. New Haven and London: Academic Press.
Dunbar, R.I.M. 2009. Evolution in Anthropology: A Comment on Nettle's "Beyond Nature versus Culture". *Journal of the Royal Anthropological Institute* 15 (2): 245.
Durkheim, Émile. 1893. In *The Division of Labor in Society. [2013]*, ed. Steven Lukes. Basingstoke: Palgrave Macmillan.
Durkheim, Emile. 1897. *Suicide: A Study in Sociology. [1951]*. New York: The Free Press.
Ehrenzweig, Albert A. 1971. *Psychoanalytic Jurisprudence: On Ethics, Aesthetics, and "Law"—On Crime, Tort and Procedure*. Leiden, Dobbs Ferry, NJ: A.W. Sijthoff – Oceana Publications.
Elster, Jon. 1985. *Sour Grapes: Studies in the Subversion of Rationality*. Cambridge: Cambridge University Press.
Emerson, Richard M. 1976. Social Exchange Theory. *Annual Review of Sociology* 2: 335–362.
Evans, Jonathan St.B.T. 2008. Dual-Processing Accounts of Reasoning, Judgment, and Social Cognition. *Annual Review of Psychology* 59 (January): 255–278.
Favretto, Anna Rosa. 2016. Come si apprendono le norme? La socializzazione normativa e giuridica. In *Lineamenti di sociologia del diritto*, ed. Amedeo Cottino. Turin: Zanichelli.
Fehr, Ernst, and Simon Gächter. 2002. Altruistic Punishment in Humans. *Nature* 415 (6868): 137–140.
Ferrari, Luigi, and Salvatore Randisi. 2011. *Psicologia fiscale: illusioni e decisioni dei contribuenti*. Milan: Raffaello Cortina.
Ferri, Enrico. 1925. Prefazione. In *La psicologia giudiziaria*, ed. Altavilla Enrico. Turin: UTET.
Fikentscher, Wolfgang. 1999. Law-Related Education, Moral Development, and the Sense of Justice. In *Law & Evolutionary Biology*, ed. Lawrence A. Frolik, Wolfgang Fikentscher, and Gerti Dieker. Portola Valley: Gruter Institute for Law and Behavioral Research.
Finkel, Norman J. 2000. But It's Not Fair! Commonsense Notions of Unfairness. *Psychology, Public Policy and Law* 6 (4): 898–952.
Fittipaldi, Edoardo. 2012. *Psicologia giuridica e realismo: Leon Petrażycki*. Milan: Edizioni universitarie di lettere economia diritto.
Fodor, Jerry A. 1983. *The Modularity of Mind*. Cambridge, MA, London: The MIT Press.
Folger, Robert. 1984a. Emerging Issues in the Social Psychology of Justice. In *The Sense of Injustice: Social Psychological Perspectives*, ed. Robert Folger. New York, London: Plenum Press.

———. 1984b. *The Sense of Injustice. Social Psychological Perspectives*. New York, London: Plenum Press.

Frank, Jerome. 1949. *Law and the Modern Mind*. London: Stevens & Sons Limited.

Frank, Robert H., and Philip J. Cook. 1995. *The Winner-Take-All-Society: Why the Few at the Top Get So Much More Than the Rest of Us*. New York: Penguin Books.

Gergen, Kenneth J. 1985. The Social Constructionist Model in Modern Psychology. *American Psychologist* 40 (3): 266–275.

Ghezzi, Morris Lorenzo. 1996. *Diversità e pluralismo: La Sociologia del diritto penale nello studio di devianza e criminalità*. Milan: Raffaello Cortina.

Gire, James T. 1997. The Varying Effect of Individualism-Collectivism on Preference for Methods of Conflict Resolution. *Canadian Journal of Behavioural Science* 29 (1): 38–43.

Goffman, Erving. 1959. *The Presentation of Self in Everyday Life*. Garden City: Doubleday & Co.

Goodenough, Oliver R. 2006. Can Cognitive Neuroscience Make Psychology a Foundational Discipline for the Study of Law? In *Law and Psychology*, ed. Belinda Brooks-Gordon and Michael Freeman. Oxford: Oxford University Press.

Gulotta, Guglielmo, and Antonietta Curci, eds. 2010. *Mente, società e diritto*. Milan: Giuffrè.

Guthrie, Chris, Jeffrey J. Rachlinski, and Andrew J. Wistrich. 2001. Inside the Judicial Mind. *Cornell Law Review* 86 (4): 777–830.

Hampshire, Stuart. 2000. *Conflict Is Justice*. Princeton: Princeton University Press.

Hensler, Deborah. 2002. Suppose It's Not True: Challenging Mediation Ideology. *Journal of Dispute Resolution* 19: 81–99.

Herrmann-Pillath, Carsten. 1994. Evolutionary Rationality, "Homo Economicus," and the Foundations of Social Order. *Journal of Social and Evolutionary Systems* 17 (1): 41–70.

Hibberd, Fiona J. 2005. *Unfolding Social Constructionism*. New York: Springer.

Hodgson, Geoffrey M., and Thorbjørn Knudsen. 2010. *Darwin's Conjecture: The Search for General Principles of Social & Economic Evolution*. Chicago, London: University of Chicago Press.

Hogg, Michael A., and Graham M. Vaughan. 2011. *Social Psychology*. Edinburgh: Pearson Education.

Homans, George C. 1961. *Social Behavior: Its Elementary Forms*. New York, Burlingame: Harcourt, Brace & World.

Houk, Carole S., and Lauren M. Edelstein. 2008. Beyond Apology to Early Non-judicial Resolution: The Medicom Program as a Patient Safety-Focused Alternative to Malpractice Litigation. *Hamline Journal of Public Law and Policy* 29: 411–422.

Jhering, Rudolf von. 1913. In *Law as a Means to An End*, ed. Isaac Husik. Boston: The Boston Book Company.

Justickis, Viktoras. 2008. Does the Law Use Even a Small Proportion of What Legal Psychology Has to Offer? In *Psychology and Law: Bridging the Gap*, ed. David Canter and Rita Žukauskienė. Aldershot-Burlington: Ashgate.

Kahneman, Daniel. 2011. *Thinking, Fast and Slow*. New York: Farrar, Straus and Giroux.

Kaplow, Louis, and Steven Shavell. 2001. Fairness Versus Welfare. *Harvard Law Review* 114: 961–1388.

Krebs, Dennis. 1982. Prosocial Behavior, Equity and Justice. In *Equity and Justice in Social Behavior*, ed. Jerald Greenberg and Ronald L. Cohen. New Haven and London: Academic Press.

Lerner, Melvin J. 1980. *The Belief in a Just World. A Fundamental Delusion*. New York: Plenum Press.

Leung, Kwok, and E. Allan Lind. 1986. Procedural Justice and Culture: Effects of Culture, Gender, and Investigator Status on Procedural Preferences. *Journal of Personality and Social Psychology* 50 (6): 1134–1140.

Levine, James P. 1975. The Ineffectiveness of Adding Police to Prevent Crime. *Public Policy* 23: 523–545.

Lopreato, Joseph, and F. Penny Anthon Green. 1990. The Evolutionary Foundations of Revolution. In *Sociobiology and Conflict: Evolutionary Perspectives on Competition, Cooperation, Violence and Warfare*, ed. Johan M.G. van der Dennen and Vincent S.E. Falger. London, New York, Tokyo, Melbourne, Madras: Chapman and Hall.

Lyon, Margot L. 1994. Emotion as Mediator of Somatic and Social Processes: The Example of Respiration. In *Social Perspectives on Emotion*, ed. William M. Wentworth and John Ryan. Bingley: Emerald.

Major, Brenda, and Kay Deaux. 1982. Individual Differences in Justice Behavior. In *Equity and Justice in Social Behavior*, ed. Jerald Greenberg and Ronald L. Cohen. New York, London: Academic Press.

Marraffa, Massimo. 2011. Evoluzione, cognizione e cultura. In *Neuroetica: Scienze del cervello, filosofia e libero arbitrio*, ed. Andrea Lavazza and Giuseppe Sartori. Bologna: Il Mulino.

McClintock, Charles G., and Linda J. Keil. 1982. Equity and Social Exchange. In *Equity and Justice in Social Behavior*, ed. Jerald Greenberg and Ronald L. Cohen. New Haven and London: Academic Press.

McEwan, Jenny. 2006. Breaking Down the Barriers. In *Law and Psychology*, ed. Belinda Brooks-Gordon and Michael Freeman. Oxford: Oxford University Press.

Merton, Robert King. 1968. *Social Theory and Social Structure*. New York: The Free Press.

Milgram, Stanley. 1963. Behavioral Study of Obedience. *Journal of Abnormal and Social Psychology* 67: 371–378.

Miller, Joan G. 2001. The Cultural Grounding of Social Psychological Theory. In *Blackwell Handbook of Social Psychology: Intraindividual Processes*, ed. Abraham Tesser and Norbert Schwarz. Malden, MA: Blackwell Publishers Inc.

Nettle, Daniel. 2009. Beyond Nature Versus Culture: Cultural Variation as an Evolved Characteristic. *Journal of the Royal Anthropological Institute* 15 (2): 223–240.

Nowak, Martin A., Karen M. Page, and K. Sigmund. 2000. Fairness Versus Reason in the Ultimatum Game. *Science* 289 (September): 1773–1775.

Oneroso Di Lisa, Fiorangela. 1989. *Giustizia, norme e autoregolazione in psicologia sociale*. Naples: Liguori Editori.

Palmonari, Augusto, and Nicoletta Cavazza. 2012. Introduzione. In *Ricerche e protagonisti della psicologia sociale*, ed. Augusto Palmonari and Nicoletta Cavazza. Bologna: Il Mulino.

Pattaro, Enrico. 2005. The Law and the Right: A Reappraisal of the Reality That Ought to Be. In *A Treatise*, ed. Enrico Pattaro. Dordrecht: Springer.

Pavlov, Ivan. 1928. In *Lectures on Conditioned Reflexes: Twenty-Five Years of Objective Study of the Higher Nervous Activity (Behaviour) of Animals*, ed. W. Horsley Gantt. New York: International Publishers.

Petrażycki, Leon. 2011. In *Law and Morality*, ed. Javier A. Treviño. New Brunswick, London: Transaction Publishers.

Pinker, Steven. 2011. *The Better Angels of Our Nature: Why Violence Has Declined*. New York: Penguin.

Poon, June M.L. 2012. Distributive Justice, Procedural Justice, Affective Commitment, and Turnover Intention: A Mediation-Moderation Framework. *Journal of Applied Social Psychology* 42 (6): 1505–1532.

Proctor, Darby, Rebecca A. Williamson, Frans B.M. de Waal, and Sarah F. Brosnan. 2013. Chimpanzees Play the Ultimatum Game. *Proceedings of the National Academy of Sciences of the United States of America* 110 (6): 2070–2075.

Quadrio, Assunto, and Gaetano De Leo. 1995. *Manuale di psicologia giuridica*. Milan: Edizioni universitarie di lettere economia diritto.

Rachlinski, Jeffrey J. 2000a. Heuristics and Biases in the Court: Ignorance or Adaptation? *Oregon Law Review* 79: 61.

———. 2000b. The "New" Law and Psychology: A Reply to Critics, Skeptics, and Cautious Supporters. *Cornell Law Review* 85: 739–766.

Ravenna, Marcella. 2012. Stanley Milgram: le ricerche sulla sottomissione all'autorità. In *Ricerche e protagonisti della psicologia sociale*, ed. Augusto Palmonari and Nicoletta Cavazza. Bologna: Il Mulino.

Reis, Harry T. 1984. The Multidimensionality of Justice. In *The Sense of Injustice: Social Psychological Perspectives*, ed. Robert Folger. New York, London: Plenum Press.

Robinson, Paul H., Robert Kurzban, and Owen D. Jones. 2007. The Origins of Shared Intuitions of Justice. *Vanderbilt Law Review* 60 (3): 1633.

Runciman, Walter G. 1966. *Relative Deprivation and Social Justice: A Study of Attitudes to Social Inequality in Twentieth-Century England*. London: Routledge & Kegan.

———. 2009. Not Whether but How: A Comment on Nettle's "Beyond Nature versus Culture". *Journal of the Royal Anthropological Institute* 15 (2): 241–243.

Sacchetti, Francesca. 2011. Sé come un altro: l'etica della reciprocità nel pensiero di Paul Ricoeur. *Societàmutamentopolitica* 2 (4): 317–345.

Sampson, Edward E. 1983. *Justice and the Critique of Pure Psychology*. New York, London: Plenum Press.

Schonewille, Manon, and Fred Schonewille. 2014. *The Variegated Landscape of Mediation: A Comparative Study of Mediation Regulation and Practices in Europe and the World*. The Hague: Eleven International Publishing.

Schubert, Glendon. 1965. *The Judicial Mind: The Attitudes and Ideologies of Supreme Court Justices—1946–1963*. Evanston: Northwestern University Press.

———. 1974. *The Judicial Mind Revisited: Psychometric Analysis of Supreme Court Ideology*. Oxford: Oxford University Press.

Schwarz, Bohner, and Norbert Gerd. 2001. The Construction of Attitudes. In *Blackwell Handbook of Social Psychology: Intraindividual Processes*, ed. Abraham Tesser and Norbert Schwarz. Malden, MA: Blackwell Publishers Inc.

Shelby Rosette, Ashleigh, Jeanne M. Brett, Zoe Barsness, and Anne L. Lytle. 2012. When Cultures Clash Electronically: The Impact of E-Mail and Culture on Negotiation Behavior. *Journal of Cross-Cultural Psychology* 43 (4): 628–643.

Sheppard, Blair H. 1985. Justice Is No Simple Matter: Case for Elaborating Our Model of Procedural Fairness. *Journal of Personality and Social Psychology* 49 (4): 953.

Sheppard, Blair H., David M. Saunders, and John W. Minton. 1988. Procedural Justice From the Third-Party Perspective. *Journal of Personality and Social Psychology* 54 (4): 629–637.

Shestowsky, Donna. 2004. Procedural Preferences in Alternative Dispute Resolution: A Closer, Modern Look at an Old Idea. *Psychology, Public Policy and Law* 10 (3): 211.

Stouffer, Samuel A., Edward A. Suchman, Leland C. DeVinney, Shirley A. Star, and Robin R. Williams Jr. 1949. *The American Soldier. Adjustment During Army Life*. Princeton: Princeton University Press.

Tanford, J. Alexander. 1990. The Limits of a Scientific Jurisprudence: The Supreme Court and Psychology. *Indiana Law Journal* 66: 137.

Taruffo, Michele. 1981. Introduzione. In *Il giusto processo*, ed. John Thibaut and Laurens Walker. Milan: Giuffrè.

Tesser, Abraham, and Norbert Schwarz. 2001. In *Blackwell Handbook of Social Psychology: Intraindividual Processes*, ed. Abraham Tesser and Norbert Schwarz. Malden, MA: Blackwell Publishers Inc.

Thibaut, John, and Laurens Walker. 1975. *Procedural Justice: A Psychological Analysis*. Hillsdale, NJ: Lawrence Erlbaum Associates.

Thibaut, John, Laurens Walker, Stephen LaTour, and Pauline Houlden. 1974. Procedural Justice as Fairness. *Stanford Law Review* 26: 1271.

Törnblom, Kjell. 1992. The Social Psychology of Distributive Justice. In *Justice. Interdisciplinary Perspectives*, ed. Klaus R. Scherer. Cambridge, UK: Cambridge University Press.

Tyler, Tom R. 2002. Procedure or Result: What Do Disputants Want from Legal Authorities? In *A Handbook of Dispute Resolution: ADR in Action*, ed. Karl J. Mackie. London, New York: Routledge.

Tyler, Tom R., and E. Allan Lind. 2001. Procedural Justice. In *Handbook of Justice Research in Law*, ed. Joseph Sanders and V. Lee Hamilton. New York, Boston, Dordrecht, London, Moscow: Kluwer Academic-Plenum Publishers.

van den Bos, Kees. 1999. What Are We Talking About When We Talk About No-Voice Procedures? On the Psychology of the Fair Outcome Effect. *Journal of Experimental Social Psychology* 35 (6): 560–577.

van den Bos, Kees, Riël Vermunt, and Henk A.M. Wilke. 1997. Procedural and Distributive Justice: What Is Fair Depends More on What Comes First than on What Comes Next. *Journal of Personality and Social Psychology* 72 (1): 95.

Vermunt, Riël. 2014. *The Good, the Bad, and the Just: How Modern Men Shape Their World*. Farnham, Burlington: Ashgate.

Vidmar, Niel. 1997. Procedural Justice and Alternative Dispute Resolution - Oñati International Series in Law and Society. In *Procedural Justice*, ed. Klaus F. Röhl and Stefan Machura. Dartmouth: Ashgate.

Wegner, Daniel M. 1982. Justice and the Awareness of Social Entities. In *Equity and Justice in Social Behavior*, ed. Jerald Greenberg and Ronald L. Cohen. New York, London: Academic Press.

West-Newman, Catherine Lane. 2005. Feeling for Justice? Rights, Laws, and Cultural Contexts. *Law and Social Inquiry* 30 (2): 305–335.

Winkielman, Piotr, Gary G. Berntson, and John T. Cacioppo. 2001. The Psychophysiological Perspective on the Social Mind. In *Blackwell Handbook of Social Psychology: Intraindividual Processes*, ed. Abraham Tesser and Norbert Schwarz. Malden, MA: Blackwell Publishers Inc.

Wright, Robert. 1995. *The Moral Animal: Why We Are the Way We Are*. New York: Vintage.

Zimring, Franklin E., Jeffrey Fagan, and David T. Johnson. 2010. Executions, Deterrence, and Homicide: A Tale of Two Cities. *Journal of Empirical Legal Studies* 7 (1): 1–29.

Chapter 3
Nature, Evolution, and Law

3.1 Sociological Theories of Human Behavior

If we can draw a preliminary conclusion from the previous chapters, it is that sociological theory and the social sciences in general have not yet articulated or proposed a convincing and organic theory of the motives of human behavior, a theory by which to adequately explain the legal phenomenon. As Gallino (1982, p. 99) has argued, this is because the social sciences have until recently been deprived of a theory of social behavior, stuck as they were in the dogma of social individuality conceived as a *blank slate*, and in the extremist positions of culturalism.

When the exchange of ideas with other branches of the social sciences, which began to interbreed their paradigms with those of the natural sciences, risked contaminating the purity (or self-referentiality) of theoretical constructions, sociology was quick to retreat to its positions and to assert the incompatibility of languages. This first happened in social psychology and sociobiology and then in developmental psychology and behavioral neuroscience, when a common framework was identified on which basis the advances that biology and biochemistry had made in human physiology could be integrated into the theory of human behavior. The proliferation of new "hybrid" disciplines (behavioral ethology, evolutionary biology, cognitive psychology) testifies to the attempts to frame the development of new interdisciplinary methodologies within a yet-to-be-established academic field.

Even in the social sciences, the is no lack of voices that are favorable to transdisciplinarity and have thus pointed out the methodological impropriety and strategic myopia of this retreat (van den Berghe 1990; Crippen 1994; Runciman 2005; Hopcroft 2009; Sarti 2015). Moreover, in the sociological insights of the discipline's founding fathers, the points of support for reviewing social behavior in a naturalistically updated fashion would be well represented. Thus, for instance, Simmel's discoveries about love and the basic emotional instincts (Simmel 1927, chap. 1.4) could be said to form the basis of the modern biochemistry of emotions, and Simmel would in turn benefit from the latter.

L. Cominelli, *Cognition of the Law*,
https://doi.org/10.1007/978-3-319-89348-8_3

Durkheim's passages on the feeling of revenge (Durkheim 1893, p. 1421), the sacredness of afflictive penal norms (ibid., pp. 1629–1638), and the heritability of tastes and attitudes (ibid., pp. 4900, 5093) seem to presuppose a knowledge of social psychology that for historical reasons would not have been possible, and they call into question the entrenched image of Durkheim as the custodian of the complete preeminence of ideology and culture and of the exclusion of psychology in the explanation of any social phenomenon. The early Marx, still far from dialectical materialism, nourished an authentic devotion to Darwinian evolutionism, recognizing its supreme authority in the explanation of the biological. He was willing to recognize the need for a unitary science of man and nature, even if this integration was more historical than naturalistic. Only later did his enthusiasm cool off, and his followers will develop his thought in an idealistic sense (Gennaro 2010, p. 15).

There is truth to Spencer's harsh judgment of dialectical materialism as a theory translated into a metaphysical idealism that, assuming a world governed by immutable universal laws, stands in irreducible contrast to the modern sciences, whose advancements it ignores. Marx and Durkheim, as well as Weber, are guided in their work by a search for the Great Idea, understood as the Fundamental Law of the Society, which could not yet be founded on the still primitive sciences of ethology and evolutionism. The evolutionist principle was organically incorporated in a theoretical *summa sociologica* in Spencer's *Principles of Sociology* (1898), but in a strongly causative and deterministic sense. Evolutionary mechanicism was then framed as the foundation of social Darwinism, thus justifying classism among individuals and imperialism among states (Collins 1975, p. 93), and it would end up being condemned by the tribunal of history and men in the Nuremberg trials.

Yet a reading of Spencerian evolutionism as a "philosophy of freedom" applied to the legal dimension had been proposed by Guglielmo Salvadori (1904, p. 15) in a theory that precedes the evolutionary analysis of law. Salvadori captures psychological individualism in Spencer, underscoring its importance relative to biological mechanicism, and noting that the Spencerian system was a direct consequence of a conception of social organization as "active cooperation, solidarity, mutual integration, and the coordination of parts among one another and with the organic unity that they form" (Salvadori 1904, p. 16).

Nor is it possible to ignore the influence that Darwinian human ethology (Darwin 1872) had on the Freudian theory of impulses, and it is worth asking why the former, unlike the latter, is not in itself considered as falling within the purview of sociology. Collins (1975, p. 90) notes that Freud, like Durkheim, implicitly supported many of the hypotheses on the animal nature of humans, even though such animality was complemented by the special ability to incorporate the social world into the mind. Freud expressly described himself as a Darwinian, and with the doctrine of natural selection he shared a disenchanted attitude toward human behavior, having identified a series of impulses which evidently had an adaptive purpose (Wright 1995, pp. 314–315, 323), and with respect to which the mind was identified as the locus of a mediation with social reality (ibid., p. 321).

A similar interpretation was proposed for the so-called symbolic-normative order, whose task was to channel the pulsions in a way that was not destructive to the individual and the community (Crespi 1989, p. 117). In human action, microsociology and interactionism highlight the presence of impulses and sociophysiological needs (Mead 1972, p. 228ff.) and of psychobiological attributes (Goffman 1969). Runciman (2005) suggested that a *stone age sociology* would be able to fill the gap and indicate, for example, the differences between acquired and imposed behaviors. Others (Machalek and Martin 2004; Maryanski 1994) have called for an "evolutionary sociology."

It should be noted that in the social sciences concerned with human behavior, *biological* is often improperly used to mean "physiological" or "genetic," giving rise to the contraposition between "the biological" and "the social." In biology, too, it is believed that social behavior is too plastic and subject to environmental variations to be studied with sufficient precision (Alexander 1986, p. 20). The biological sciences therefore confine themselves to the study of nonhuman behaviors or to behaviors that can be directly traced to genetic or physiological causes, and certainly not to behaviors involving aspects considered exclusive to humans, such as politics and culture. Alexander (1986, p. 20) incidentally observes, with some amazement, that it is possible to be awarded a doctorate in the social sciences without having the slightest idea of contemporary advances, even the most basic ones, in the natural and biological sciences.

The hegemony that constructivism holds over the social sciences has nevertheless kept at a safe distance any hypothesis that might contribute to explaining the social from a biological point of view. Encroachments on the field have been rejected through the peremptory argument that to grant the validity of such explanations is to legitimize social Darwinism and ultimately, with a facile *reductio ad Hitlerum*, Nazi eugenics. When biological-naturalistic paradigms are mentioned, as in Foucault's (2004) biopolitics, it is to mock them as the dependent variable in political-hegemonic processes, and within a hyper-structuralist interpretation that in any case denies that falsificationist science can have any objective value. Today, skeptical positions have found new expression in criticism aimed at the "naturalistic turn" and laying the charge of "scientific imperialism" (Mäki et al. 2018).

This anti-positivist objection to the commingling of biology and sociology carries antimodern implications whose full import we will see later on in Sect. 3.5, when we turn to the "battle for science" in the debate on what human nature is (Chomsky and Foucault 2006, pp. 138–139). As Antonio Pennisi (2014, p. 12) reminds us, however, biopolitical theories fall into the old trap of *Plato's error*, which consists in pretending that we can "engineer the political governability of the minds and bodies of human aggregates." This pretense is the outcome of Platonic constructivist utopianism (Popper 1971). In opposition to these Foucauldian biopolitics, Pennisi sets an "Aristotelian-Darwinian biopolitics," on which basis the political animal, placed in the "theater of natural selection and equipped only with language," pursues reproduction and movement in the surrounding environment (ibid., p. 7).

Few old-school sociologists, apart from Runciman and Elias, have offered a sociological criticism of the hegemony the constructionist paradigm wields in the social sciences, and with it, of the hostility toward any attempt to affirm the relevance of the evolutionist principle in the study of human behavior and sociality. Sociologists have attempted to expunge the biological element even from fields such as demography and ecology. Here it bears mentioning Van den Berghe, for he was among the first researchers to see that anthropology has come to the aid of sociology, so far with poor results, by providing insights with which to better understand adaptation to the habitat. Van den Berghe (1990, pp. 174–176) speculates about the different reasons for this resistance. The difficulty of reasoning in evolutionary terms in sociology may primarily be the result of our humanity. Evolutionary theory makes it necessary to ponder phenomena unfolding on a timescale that is inconceivable for the individual human mind and experience. Even history cannot look beyond arcs of time that, while historically eventful, do not necessarily see significant biological variations in evolutionary terms. Evolutionary theory, then, presupposes the occurrence of mutations which spread across the population because adaptively advantageous, but whose origin remains random. This randomness sits poorly with the intentionality that we intuitively project onto the world around us.

There is widespread generic anthropocentrism in all individuals, which leads us to consider ourselves unique and different from all other living beings. Finally, we incline toward a cynical view of behavior, which sees every altruistic gesture as prompted by nepotism toward one's family or allies, or else as carrying a demand for reciprocity. For a clear example of an anthropocentric ethological conception that has now been disavowed by scientific observations of animal behavior, we can turn to Jhering (1913, p. 181): "Life of the stronger at the expense of the weaker, annihilation of the latter in conflict with the former,—such is the form of life in the animal world; assured existence also of the weakest and the poorest by the side of the strongest and mightiest,—such is the form of life in the human world."

Directly linked to the status of social scientists is the cultural anthropocentrism that derives from being a professional sociologist. Here, the sociologist would deliberately choose to ignore the biological standpoint, because scientifically irrelevant and perhaps dangerous. Even the so-called biosociological currents, while conceding that evolutionary dynamics can play a role in explaining human behavior, adhere to the dogma of the uniqueness of human beings, arguing that the analogies with the world of nonhuman animals would merely amount to anecdotes lacking any general explicative value.

The "new orthodoxy of sociocultural environmentalism" (van den Berghe 1990, pp. 177–178) is the lowest common denominator of sociological schools that revolve around the false dichotomies of nature/culture, heritability/environment, instinct/learning. Van den Berghe's polemical argument goes so far as to accuse sociologists of merely using data (often collected by others) that have been aggregated into abstract categories, and then manipulating the data to establish correlations between variables that are too remote from concrete social behaviors (van den Berghe 1990, p. 181). No less contentious is Runciman (1999, p. 57), who sees the social sciences as disciplines peopled by *merchants of platitudes*—so sure of certain general truths about behavior as to be convinced that reformulating them into high-sounding formulas can contribute to sociological knowledge—and *merchants of attitudes*, which fraudulently smuggle a personal political view by ignoring the distinction between facts and values.

The clarification that would derive from a unified theoretical paradigm of individual behavior is unacceptable to these sociological approaches, which have epistemologically coalesced and crystallized into the Standard Social Science Model (Tooby and Cosmides 1992, p. 22). To subscribe to this model is to live in the illusion that the intellectual uniqueness of human beings makes them to some extent unpredictable, which in turn necessarily complicates the schemes by which to explain human behavior (van den Berghe 1990, p. 182). I would not be as optimistic as Runciman in believing that we are done with the most dogmatic simplifications of Marxism, social Darwinism, behaviorism, structuralism, and Durkheimian cultural anthropology and with postmodernist oversimplifications. Because of the stigma of social Darwinism, many sociologists are still reluctant to look to biological Darwinism to derive useful elements for their sociological theory. Yet there are no methodological grounds on which to reject the biological study of social behavior, nor does such a study have to be Darwinist. Both intrapersonal-psychological and interpersonal-sociological behavior are "effectors of organic apparatuses integrated in a systematic psychosomatic unity, namely, the organism" (Bertacchini and Genta 1982, p. 125).

An interesting proposal, based on some elementary assumptions of evolutionary psychology, was formulated to replace the Standard Social Science Model with an Integrated Causal Model (Tooby and Cosmides 1992, pp. 23–24). The following propositions illustrate the Integrated Causal Model: the human mind consists of a set of systems for processing information that are rooted in the nervous system; these mechanisms are each specialized in resolving a different adaptation problem through language, partner choice, family relationships, and social cooperation; the same information-analysis mechanisms generate what we identify as human culture (behaviors, artefacts, linguistic representations); the cultural contents generated through these mechanisms are adopted or modified through psychological mechanisms shared with other members of the population; these dynamics activate epidemiological and historical processes to rebalance the population; these processes are not always generalizable, because they are set in particular ecological, economic, and demographic contexts or environments.

In my view, it is precisely from this evolutionary psychology that we can derive the most stimulating contributions to the social sciences. Following an approach of vertical conceptual integration, all sciences that analyze behavior should aim at mutual consistency (Laland and Brown 2011, p. 195ff.). Two basic assumptions that would sustain this view and outcome in sociology are that (1), although there is no human nature at the level of expressed cultural behaviors, there *is* one at the level of evolved psychological mechanisms, and (2) these mechanisms have emerged through natural selection on a timescale which is compatible with evolutionary dynamics, and which for human beings began in the Pleistocene epoch, when communities lived as hunter-gatherers (Barkow et al. 1992, p. 5). This process of crystallization does not mean that evolution has stopped, but that the subsequent environmental changes would still be too recent and sketchy to fundamentally modify our genome (Freese et al. 2003, p. 237). As will be made clear further on, the transition to the agricultural society, often regarded as the reference point for a past golden age, was the first significant cultural rupture in human existence, imposing social regimes and forms of life that had hitherto been "unnatural."

The analysis of law as a social system has certainly conformed to the Standard Social Science Model, only borrowing from philosophy and literature when needed, and resting content with a teleological or poetic vision of human nature. The observation that moral intuitions should generally always be subjected to "suspicious scrutiny" in behavioral analysis appears very acute (Alcock 2001, p. 205). Behavioral models have rarely been made explicit, because they have played a marginal role within a constructivist and systemic paradigm. It has also been pointed out that the functioning of the legal system is based on several beliefs about human nature that are often outdated or even erroneous (Jones and Goldsmith 2005, p. 421).

What complicates the analysis of legal attitudes is the diversity of legal systems, a diversity that in turn reflects profound cultural differences. Why is polygamy illegal in some countries and legal in others? Why is the buying and selling of human organs or blood in some cases legal in some systems and illegal in others? The law works in a reactive and ad hoc fashion to social emergencies, and these emergencies are fossilized in the appearance of a substantial lack of homogeneity in the natural laws that govern human behavior. If we took this line of reasoning to the extreme, we could conclude that individuals in Japan or the United Kingdom are fundamentally different from us because they drive on the left side of the road. This is just an example of arbitrary coordinated behavior, where rules of conduct are chosen on a purely conventional, if not arbitrary, basis, for what matters is that everyone should conform to the convention, and in this conformity lies its advantage to society.

The misapprehension of the social sciences, and especially of the law, is to postulate social subjects who are socially collectivistic but instinctively selfish and competitive, whereas humans, from a social point of view, tend to be individualistic but also instinctively cooperative in their interactions. In essence, when facing the macro side of society, we are critical and suspicious, while in micro-interpersonal social relationships, we incline toward collaboration, with the significant exception of manipulators, cheaters, and free-riders.

Maryanski and Turner (1992, p. 68) observe that sociology has always had a communitarian and collectivistic bias, probably on the false assumption that human beings are by nature exclusively aggressive, hierarchical, and deeply territorial. What we know today about hunter-gatherers seems to cast doubt on this inevitability, which instead is based on our understanding of agrarian societies and the centralized political aggregates that followed them: since it was only at that point that writing became widespread, it was on the basis of these written records that the first sociological theories of behavior were built, but it is contingent that our sociological knowledge of human behavior should be thus skewed. This misunderstanding is highlighted in what may be referred to as the *paradox of cooperation*. If the assumption is that we seek sociality but are intimately selfish, we should expect the legal system to require us to cooperate and be altruistic, forcing on us those choices that in the long run make for social utility. This is the so-called *paradox of collective action* (Nielsen 1994, p. 291), which constitutes the theoretical formalization of the equivalent paradox at the micro-level, namely, *altruism* (Wilson 2015).

The reality is that we do not seek sociality as a standard rule of existence, but we instead look for cooperative exchange, which is quite different, and to reinforce the latter, we are often willing to do altruistic acts that make us look disinterested. The law, for the most part, understands and elevates selfish tendencies to a standard rule of behavior. Think of the rules and duties of care, which shift to those who have been wronged the burden of assuming a generalized malevolent intent in the community, and when this duty is not fulfilled, the solution is to deem the aggrieved party jointly responsible for the injuries he or she has suffered.

3.2 The Paradox of Cooperation

I have argued that the sociological, social, and behavioral sciences must and can coordinate in a framework that is coherent with the biological sciences. In support of this view, we can rely on a bridge-law asserting the universality of cooperation as a guiding principle in the coexistence of living organisms. Our behavior is not significantly different from that of other complex living beings, and the behavior of these beings is not significantly different from that of single-celled organisms. The point here is not to reduce our own behavior down to the level of animal behavior, but to raise the behavior of nonhuman organisms by bringing out its closeness to our own (Pocar 2005).

Nonhuman ethology has amply demonstrated that the behaviors of all living organisms are incredibly more sophisticated than what was imagined even only a few years ago, when much of the sociological theory of behavior had already been consolidated. Cooperation and the tendency toward mutual exchange should, in my view, supplant the collectivistic interpretation principle (and its individualistic converse), which posits a coordination of basic behaviors for the achievement of a unitary and superior goal. As highlighted when the Standard Social Science Model was introduced, the social sciences have so far been under the sway of the dyadic collectivist/individualistic paradigm. Depending on the lens through which we choose to interpret history, human society is either born collectivistic and distortedly evolves toward individualism (Rousseau), or it is born individualistic and evolves toward forms of collective control (Hobbes). I submit that the swinging of the pendulum between individualism and social collectivism is a contingent manifestation: underneath it lies an individual human constant that continuously goes back and forth between cooperation and its obverse, competitive defection.

The concept of cooperation is common to many other levels of biological life, and not only among complex organisms. According to evolutionists, the feedback between the micro and the macro could be just the interpretive key to the emergence of a "macroscopic harmony from microscopic selfishness" (Wright 1995, p. 198). The "coalitions of genes" that constitute the genome of every living being have been included among the multiform manifestations of biological cooperation, and so have the cells that make up multicellular organisms, or the various cooperative actions that are carried out by groups of microorganisms (Foster et al. 2007, p. 74; Nettle 2009, p. 226). In the biological processes of contagion and imitation, Gabriel Tarde (1903), the theorist of vogues and fashions and unlucky rival of Durkheimian "sociological realism," identified a general analogical model of social change. The benefits of imitation and self-imitation among living beings have also been highlighted by Mormino (2016).

A key moment in the construction of the theory of cooperation came with Axelrod and Hamilton (1981), an economist and an evolutionary biologist who worked on an ambitious project in game theory that introduced the evolutionary element in cooperative strategies, seeking to determine how the relation between cooperation and competition may become evolutionarily stable through the establishment of regular and recognizable individual patterns of behavior. The experiment, which took more of a strategic approach than a genetic one, is now well known for its final results, but incredibly little known in sociological literature. The question that Axelrod specifically tried to answer was: how can the practice of cooperation emerge and stabilize in an evolutionarily unstable world without any central authority? In the *prisoner's dilemma* (Kuhn 2017; Runciman 1970, p. 225ff.), the philosophical conclusion was that it is always cheaper and less risky to betray than to trust and collaborate. As is known, if Cold War strategy had been left to the strategic suggestions of the prisoner's dilemma, the conclusion would have been to attack first, strike hard, and hope to survive.

The originality of Axelrod's experiment was to simulate the continuity of social dynamics by comparing the different strategies of cooperation/defection over a series of iterated interactions. In particular, reiterated social interactions were simulated by making individuals randomly meet other individuals with different strategies, each time for at least 200 times, and calculating each strategy's final outcomes on the basis of conventional criteria for assigning a score, similar to those of the prisoner dilemma. The maximum gain was for the defector who met a cooperator, and the maximum loss was for the cooperator who met the defector, while moderate gains and losses were respectively assigned to the cooperators and defectors interacting with others who were using the same strategy.

The outcome could not have been any clearer: the most profitable strategy—the one that tended to establish itself as prevalent, even without eliminating other types of strategies—was the one referred to as *tit for tat*. The tit for tat could be summarized in this simple choice: always begin an interaction in a cooperative way, and then reciprocate whatever move is made by the other. Thus, in case of the other's defection, the tit-for-tat player would shift to defection, but in case of cooperation, he or she would immediately return to cooperation at the next move (Axelrod 1984, p. 31).

The properties that characterize winning strategies in a long series of repeated interactions are summarized by Axelrod as follows: avoid unnecessary conflicts and cooperate as long as your neighbor cooperates; immediately react to unprovoked defections; even after you have reacted, forgive and return to cooperation if there is a cooperative response; maintain a coherent behavior, so that your neighbor can adapt and regulate autonomously (Axelrod 1984, p. 20). The winning strategy in a nutshell, was kind but retributive: ready to forgive but clear and inflexible. An unconditional willingness to cooperate would strengthen competitive strategies, enabling aggressive tactics to almost invariably prevail. Excessive reactivity would instead bring on endless retaliation. The social environment has once more been recently described as one of "social dilemmas" (Friedman and Sinervo 2016, p. 337ff.). Social balance between individuals seems therefore to be based on "antagonistic cooperation" (Lopreato and Green 1990). Axelrod goes so far as to say that an even more fruitful strategy could be a slightly less reactive version of *tit for tat*: a strategy that would shift from cooperation to defection after two defections (Axelrod 1984, p. 39).

In individual terms, the best scores were obtained by adopting other, more aggressive strategies set up to simulate deception, like those based on this set of instructions: cooperate only once every three moves, or cooperate but at the first defection begin to defect and continue along this course until the end. In a social context of repeated interactions, excessive competitiveness was only harmful, because a niche of cooperative individuals would in any case raise one another's average score. Among the fifteen most effective strategies in the experiment, only one tended not to have a cooperative footprint (Axelrod 1984, pp. 33–35, 44). As if to confirm the Standard Model's firm grip in spite of its explanatory inadequacy, Axelrod notes that when scholars who have been trained in political science, sociology, economics, psychology, and mathematics take part in the experiment, they tend to adopt competitive strategies: they do so following the tenets of their own social theory, and despite the fact that over the long term these strategies lead to losses.

The experiment confirms that cooperation, when based on reciprocity and long-term relationships, becomes an evolutionarily and socially stable strategy, protecting itself from the encroachment of competitive strategies. An evolutionarily stable strategy is defined as a strategy that, when adopted by the bulk of the population, cannot be displaced by other strategies and will result in punishment for those who deviate from it (Dawkins 1976, pp. 74, 75). These evolutionarily stable strategies appear to be functional to the group, but without implying any deviation from a methodologically individualistic principle (Nielsen 1994, p. 278). While *tit for tat* and other cooperative strategies do not enable adopters to *individually* fare better than under other strategies—individual scores are on average equal to or lower than those obtained with other strategies—they do elicit cooperative behavior in other individuals, enabling everybody to do well, and as a result they score cumulatively higher.

In other experiments in behavioral economics, where choices are made salient by highlighting their social impact in the medium or long term, an individual tendency to allocate goods in a fair and disinterested manner has been proved empirically (Turillo et al. 2002, p. 839). In a social environment traditionally conceived as starting from a state of war of all against all, a small group of cooperating individuals may be able to establish itself firmly and then spread through processes of imitation through learning and selection: it can do so when it has the opportunity to relate, even only partly, with others of its own kind. For "many of the benefits sought by living things are disproportionately available to cooperating groups" (Axelrod and Hamilton 1981, p. 1391). If people who have common values tend to like and reward each other (Homans 1961, p. 316), the dynamic is further reinforced when these common values make for cooperation.

Direct application of tit-for-tat strategies is feasible in legal disputes. As Gilson and Mnookin (1995) note, in several conflicts each of the parties feels obliged to make an aggressive judicial move to prevent the other from taking advantage of a signal of weakness. In a highly competitive environment, signs of a willingness to cooperate are at high risk of being misunderstood (Gilson and Mnookin 1995, pp. 192, 195–196). To this it can be added that lawyers may be tempted not to cooperate just so as to slow down the procedure and maximize their fees, even though many of them do not perceive themselves to be engaging in aggressive strategies, and indeed identify with strategies of the tit-for-tat type (cooperate by default, but retaliate if the counterparty should attack). In conclusion, the concrete result of prelitigation and litigation strategies is generally much more inefficient than a situation in which both parties are enabled to cooperate from the outset (ibid., p. 205). On the other hand, even if cooperative social interaction is established as a stable and prevailing strategy, this does not imply that relatively safe and stable "ecological niches" will no longer be available for defectors, too (Frank 1992, p. 56).

Although Axelrod's model has not been challenged, it is certainly possible to formulate some objections to the possibility of using these findings as a foundation for a general sociology of behavior. These are not the same objections that a structuralist macro-epistemology could direct at game theory. For, as noted, the theory I am working toward follows a micro- and individualistic approach. But it is doubtful that the tit-for-tat model can be generalized to any kind of social interaction. In some existential or institutional situations, for example, the price of losing is not a simple penalty but physical elimination or incarceration, and defeat is therefore certain: social interplay ceases to unfold over the long course, because there will be no other opportunities for cooperation, and in the absence of any external constraints the extreme defector will prevail.

Furthermore, in Axelrod's tournaments, the relative value of payoffs is clearly standardized and hypothetical. We can hypothesize that in a real-life situation, where the magnitude of rewards and punishments can vary greatly, this cannot make much difference to a cooperative person, but in an aggressive relationship it could determine significant changes in behavior. A competitive individual could strategically decide to have a cooperative demeanor when the payoff is low, and then behave opportunistically when the payoff is high or very high. Even if in this scheme there is no aspect explicitly labelled "familial" interaction, we can assume that even in selecting a partner, making reproductive choices, and choosing an education or course of training, we are subject to cooperative versus competitive logics (Wright 1995, p. 57ff.).

A factor of greater realism, in partial correction of the model's abstractness, may be introduced by speculating that a mechanism of kin selection is already at work at this stage, making it possible to explain the puzzling phenomenon of cooperative strategies. Cooperation is instinctively easier among members of the same species, and even more so among blood relatives. If the players are genetically close enough, the altruistic strategy can have some edge in spite of the defeat of single altruistic individuals (Axelrod 1984, p. 89). Axelrod's experiment, however, marked a remarkable advancement, showing that mutual cooperation can become relatively stable even in large populations of individuals, provided that individuals understand that (a) there may be future occasions of further interaction with the same individuals (Axelrod 1984, p. 174), and that individuals can (b) easily recognize each other, through the cognitive faculty of identifying faces (ibid., p. 102), and (3) remember how the other individual has behaved in the past, and in particular what strategies he or she adopted (ibid., p. 125).

The evolution of cooperative social models in complex organisms with evolved brain functions has largely been fostered by the appearance of mutual altruism (Wright 1995, pp. 200–201). It is reciprocity between persons in the same family that initially contributed to spreading the reciprocal altruism trait, to the point of reaching *hyper-cooperation* (Burkart et al. 2014) or even *hyper-citizenship* in political groups (Pitasi and Petroccia 2018; Pitasi and Angrisani 2013; Pitasi 2012).

The tendency to be altruistic is unfavorable from a strictly biological point of view, except when those involved are related and share the traits of altruism. Natural selection favors those who are willing to help others based on their degree of kinship and fertility, as well as on other characteristics that increase the beneficiary's ability to support the donor's *inclusive fitness*. In life-threatening situations, when individuals are faced with the dilemma of having to choose who is to be saved, they tend to favor close relatives over distant ones, younger persons over older ones, the healthy over the sick, the rich over the poor, women of child-bearing age over women who are past that age: it has been hypothesized that this is because we choose to help people who have greater reproductive potential. When instead it comes to helping someone in a situation of *low* risk, individuals give less importance to kinship, and prefer the young and the old to the adult, the sick to the healthy, and the poor to the rich (Burnstein et al. 1994). This suggests an instinctive and intuitive ability we have to distinguish decisions on the basis of their reproductive consequences.

But a simpler explanation seems to come from the self-defeating altruism of social insects: some individuals "accept" to play a role that will make them sterile within the colony, to the benefit of the reproductive wellbeing of the family broadly understood. This explanation has been provided in sociobiology (Wilson 2000). More difficult to explain is absolute sacrifice, in which we put our own lives at risk to preserve those of others (Alcock 2001, p. 180). Those who choose martyrdom adopt an optimal behavior for the reproduction and transmission of their "culture-type," to the detriment of the reproduction and transmission of their own genotype (Gianola 1995, p. 815). The classic example is that of individuals who take very high risks or face certain death in war in order to save their comrades, even if they can be sure that there is no kinship relation between them. What kind rationality might underlie this behavior? There is certainly a "genetic" rationality in that, in rough approximation, we tend to look on our soldier compatriots as not-so-distant relatives.

Kin selection, however, is not just familistic selfishness. We tend to represent society according to a dyadic scheme where each subject is either a family member or not, and on the basis of this representation, we assume that there is a distinction between family relationships and social relationships. On the other hand, we can all regard ourselves as members of a common family, on a continuum that goes from the closest family members, or relatives, to persons we do not consider family, but who share a substantial genetic makeup with us. Regardless of how long the period is that genetically separates us from others—a few years, from our parents or siblings; 40,000 years, between the Europeans and the Australian aborigines; or several million years, between humans and primates—the difference in what concerns the motivations that lead us to act in one way or another is not just qualitative but also quantitative.

In a sense, we consider relatives not only the closest family members, but also the closest social groups and communities, on the basis of apparently irrational feelings triggered by phenomena such as attachment to the flag or the homeland or, less consequentially today, the home team. The concept does not seem sociologically so distant from the idea of *fraternity*, as opposed to the friend-or-foe logic of political rationalizations (Resta 2002). Altruism and cooperation cannot be considered socially learned behaviors or a cultural manifestation. It is certainly true that we are never hostile to someone without a reason, but the reason is often simply that we have no interest in making friends with this person. Searle highlights the error of failing to problematize egoism by considering it the default attitude, and he claims that just as natural as the egoistic desire is altruistic desire (Searle 2001, p. 149).

Naturally, there is also a strategic advantage that we seek to obtain through cooperative attitudes. If we live in a society where altruism is a practice, then we can make friends, who can in turn help us in the event of social conflict (Wright 1995, p. 248). The altruistic behaviors that are more socially widespread among humans have been traced back to some fundamental types: helping in the event of danger (when someone in our own species or some other species has an accident or comes under attack); sharing food; helping the sick, the injured, the very young, or the very old; sharing tools; and sharing knowledge (Trivers 2002, pp. 33–34). These forms of reciprocal altruism evolve and take hold when social units begin to concentrate, when the size of the group enables mutual dependence, and when the lifespan of individuals begins to increase, with an accompanying increase in the period needed for child-rearing, when parental care is essential for survival.

It is necessary to return briefly to the feeling of revenge that is activated by the perception of injustice. The point was highlighted in the earlier treatment of social psychology (Sect. 2.3), when we looked at the experiments carried out through the ultimatum game (Nowak et al. 2000; Zak 2006). The point carries even greater implications when it is amplified by the reiteration of interactions, so much so as to suggest the coining of the expression "strong negative reciprocity" (Dauer 2006, p. 267). "Strong negative reciprocity" is the tendency to punish someone else's selfish behaviors, even if this comes with high costs and no reward, and without considering the possibility that there may not be any future contacts. According to Dauer, this behavior is not necessarily motivated by a desire to discourage the defector in the future, given that it is not certain that we will meet him or her again, but rather by an intention to show observers that we have a tendency to cooperate. The cost borne by the punisher in inflicting a particularly severe punishment would essentially come down to "reputational investment." Moreover, this kind of aggression is often disproportionate to the offense committed. Trivers (2002, p. 40) terms this type of behavior "moralistic aggression," hypothesizing that even injustices that would look small on an existential scale can still greatly benefit those who perpetrate them, and notably can disadvantage the entire community.

The universality of strong negative reciprocity, at least in some sectors of the animal world, seems borne out by the repetition of the ultimatum game with chimpanzees: their results closely resemble those with human beings (Brosnan and de Waal 2003). Primates, too, have very strong evolutionary reasons for developing a strong sensitivity to inequalities (Brosnan et al. 2007). In fact they cooperate habitually in hunting, defending the territory, forming coalitions, and having a long memory for past interactions (Proctor et al. 2013, p. 2). The same types of behavioral similarities between humans and chimpanzees can also be observed in the *endowment effect* (Brosnan et al. 2007), later discussed in Sect. 4.5. It is advantageous to report one's willingness to cooperate, because reputation is a form of social capital: cooperators seek out other cooperators and try to avoid free-riders (Dauer 2006, p. 271). Anger and the desire to take revenge on defectors are human characteristics (McGuire 1992, p. 32), but they also extend beyond the human. Along with disgust, anger is the main "emotion of condemnation" (Rozin et al. 2009), and it constitutes a more sophisticated system of aggression that is useful precisely in punishing the defector (Pinker 2002, p. 238).

According to the evolutionist interpretation, large-scale social interaction would have contributed to reinforcing and disseminating in the population characteristics such as a particular sensitivity to the advantages of cooperation, an attentiveness to the behavior of free-riders, a particular motivation to obey and implement social norms (Dauer 2006, p. 268), and a sense of justice (Trivers 2002, p. 16). Other recent experiments in the field of social interactions have brought to light the existence of punitive strategies which defectors put in place to the detriment of cooperators, and which are motivated by the very fact that even altruistic subjects need to be punished (Rand and Nowak 2011), perhaps just to bring their evolutionary advantage back into balance. In the experiment, different types of subjects were asked to contribute in various ways to a common good, choosing whether to play the role of cooperator, defector, or loner and then deciding which of the opponents to punish (Rand and Nowak 2011, p. 2).

This makes it necessary to rethink or adapt theoretical models that do not foresee the possibility of such behavior. In this case, punitive acts can be interpreted as an instrument of protection against potential competitors. A band of cooperative subjects who can punish defections consistently are clearly a problem for competitive subjects. In this case, the punishment would be motivated not by a sense of justice but by "antisocial" purposes, and would be aimed at protecting a particular strategic niche that could otherwise not survive (ibid., p. 2).

Ultimately, capitalist society is no more distant from human nature than are agricultural societies, with their hierarchies, stratification, and accumulation (Maryanski and Turner 1992, p. vii). The criticisms that modern sociologists have directed at the social alienation brought about by industrialization were based on an inaccurate and romanticized vision of agricultural societies (ibid., p. 167). Modern social relations have therefore been structured, in their most complex manifestations, as a self-imposed cage. This was made possible by a much larger brain structure that freed cognitive resources for learning and for modifying instincts (ibid., p. 90). According to this hypothesis, the instincts for combat, hierarchy, or territory would consist of "adaptive socio-cultural mechanisms" that are essential in overcoming the contradictions between the need for a more structured social organization and our biological attitudes (ibid., p. 111).

3.3 Sociobiology and the Epigenetic Hypothesis

A decisive but controversial step toward founding a unitary science of the behavior of the living came with the 1975 publication of Wilson's *Sociobiology* (which has since been updated and republished several times: Wilson 2000). Wilson is a biologist who has mainly dealt with social insects, and the building blocks of his theory can directly be found in the entomology and ecology of populations, where increasingly convincing homologies are found between human societies and insect colonies. Wilson recognizes a role for culture but argues that the biological approach in the description of social aggregates cannot be classified uncritically as deterministic reductionism. In the updated 2000 edition I am referring to here, Wilson points out that the legacy of sociobiology is being taken up today by less controversial disciplines, such as evolutionary psychology and evolutionary biology, which try to understand the mechanisms of social behaviors in light of evolutionary mechanisms (Wilson 2000, p. vii).

Wilson makes a first important systematic attempt to harmonize the principles of the social sciences with those of the biological sciences, stating that social organization is "the class of phenotypes furthest removed from the genes": it is determined jointly through the behavior of individuals and a population's demographic qualities, which in turn represent "highly synthetic properties"; "a small evolutionary change in the behavior pattern of individuals can be amplified into a major social effect by the expanding upward distribution of the effect into multiple facets of social life" (Wilson 2000, p. 11). In fact, behavior is that part of the phenotype that is most likely to vary when long-term changes take place in the environment (ibid., p. 13).

Each social species is based on a mix of personal survival, reproduction, and altruism. And social evolution is the result of a genetic response to ecological pressures (ibid., p. 32). Social behavior, in essence, would consist of a much more complex articulation and interpretation of the interaction between the environment and culture, but the latter would be explained by way of such a peculiar view of the former that the account can deflect accusations of reductionism: the cultural aspect would be "for the most part adaptive in a Darwinian sense" (ibid., p. 560). I will further argue that reductionism may be ultimately be superseded by the epigenetic hypothesis, which also sees environmental influences as interacting with the genome on an equal footing.

Wilson gives some indications for the social and behavioral sciences. First among these is the need to "temporarily" subtract from philosophers ethical evaluations, so that they can be adequately "biologicized" (Wilson 2000, p. 562). As for psychologists, those who adhere to a nonevolutionary approach would end up producing nebulous independent variables of a purely theoretical nature (ibid., p. 22). Sociology, now in its mature phase, is proving incapable of producing reliable conceptual systems, and confines itself to labeling phenomena and concepts, with imaginative definitions and metaphors. It doesn't do anything significantly different from what natural history does today. Social processes are also difficult to analyze because the fundamental units of analysis under consideration are elusive, if not perhaps inexistent. Wilson predicts that sociology could therefore soon merge with cultural anthropology, social psychology, and economics (ibid., p. 574) and accept the premises of neo-Darwinian evolutionary theory, where each phenomenon must also be evaluated on the basis of its adaptive meaning (ibid., p. 4). Even though an evolutionary sociobiology with this imprint does not yield hypotheses that can be quickly tested with elegant laboratory experiments, it could help to bridge the gap between the social and biological sciences.

As could be expected, Wilson's sociobiology set off a firestorm in the scientific world of the 1970s. Although in some respects it could neither be considered a new discipline nor a revolutionary science (Alcock 2001, p. 19), but rather an intellectual approach to a given set of problems (Elliott 1985, p. 73), its clear reductionist implications were deemed unacceptable, especially from a political point of view, because according to their detractors they were going to implicitly legitimize relations of power and domination in society. Among the most ardent critics were also some famous biologists who defended a strict separation between the biological and social sciences, especially on an ethical level. Segerstråle (2000, p. 204ff.) thoroughly documented the scientific disagreement between the well-known biologist Lewontin and Chomsky. Chomsky, a libertarian with politically radical ideas, as well as various proponents of sociobiology and the evolutionist approach, refuted the thesis that postulating the existence of a biological human nature (subject to evolutionary mechanisms not dissimilar from those of other living beings) would imply a legitimization of the current political and economic regime.

Of course, even the social sciences did not stand passive: some condemned Wilson's "genetic imperialism," and in particular the prediction that sociology would soon be absorbed into sociobiology (Rossetti 1982, p. 119), as well as Wilson's vision of a "reductionist-synthetic" science (Ingrosso and Manghi 1982, p. 20ff.). In its relationship with evolutionism, contemporary sociology is certainly more attracted to the Darwinian-anthropological than to the Darwinian-psychological version (Nielsen 1994, p. 294): the former emphasizes phenotype plasticity and its compatibility with the theories of rational choice, because it is less linked to the biological factor; the latter ends up being considered close to a Paretian model that is now outdated. Other sociologists, even if they did not fully subscribe to the sociobiological hypothesis, appreciated the systematic and synthetic effort, while also deeming it "no longer sustainable to assert a discontinuity between humans and other living species" (Ciacci 1983, p. 29). Dennett (1996, pp. 453, 461) reckons Hobbes and Nietzsche among the proto-sociobiologists. It was also argued, with regard to Parsonsian functionalism, that it could have overcome some complex theoretical obstacles if worked into a broader theoretical hypothesis of an ecological or evolutionary type (Addario 1994, p. 124). The Parsonsian willingness to embrace an evolutionary perspective are also highlighted by Vignera (2010, p. 182), who accepts the concept of the "evolutionary universal" as a complex of connected structures and processes whose development increases the ability of the living to adapt.

Gallino set out a research agenda based on four fundamental social interests which inform biological actions, and which consist of an evolutionary but not genetic reformulation of the sociobiological hypothesis. Specifically, the four interests posited in Gallino's agenda consisted in (1) a punctuated reproduction of one's self, instant by instant, maintaining a structure and identity; (2) the preservation of one's cultural structures; (3) the multiplication of one's biological replicas through offspring; and (4) the reproduction of one's experiential and cultural structures among other types of descendants, including those who are not consanguineous (Gallino 1982, p. 103). Later, Gallino defended the sociological potentialities of the evolutionary theoretical apparatus used for describing ultimate individual goals, namely, survival, persistence, and replication (Gallino 1987, p. 49). While recognizing the importance and fruitfulness of the transdisciplinary sociobiological perspective in the study of social behavior, Parisi spoke of a "reductive sociobiology" that swings between two extremes: "it either exaggerates the importance accorded to the individual as a fundamental benchmark for the effectiveness of evolutionary strategies, or it magnifies the importance of genes and regards the individual as merely their ephemeral container" (Parisi 1982, pp. 97, 85).

However, the contribution of sociobiology or, stated otherwise, the attempt of the biological sciences to establish a bridgehead in the social sciences, finds many other areas of engagement: in addition to the theory of collective action and the paradox of cooperation, ideas of sociobiological lineage produce convincing paradigms even in the sociology of gender (Nielsen 1994; Rossi 1984). Considering that sociobiology has not yet produced a very well-developed body of knowledge, it is easy to object to the way sociobiologists make use of carefully selected comparisons between human and animal behavior. On a purely speculative basis, they often produce *post-factum explanations* and *just-so stories* that are difficult to test (Alcock 2001, p. 5; Wright 1995, p. 42). Many of the correct insights that human sociobiology generates within the social sciences are thus dismissed because of their empirical failings, and because of what has been described as a *dilettante* approach (Laland and Brown 2011, p. 71).

To the sociobiological hypothesis, we can then add the suggestive theory of the *selfish gene* (Dawkins 1976). The theoretical recombination of this theory of "moral ethology" with sociobiological theory effects a mutual reinforcement and raises new questions for the social sciences. Dawkins's provocative assertion is that "individuals are survival machines programmed by their genes," since genes "exert ultimate power over behavior" (Dawkins 1976, pp. 63, 64). In order to explain the perception of acting by means of an individual consciousness, Dawkins hypothesizes that such consciousness emerges "when the brain's simulation of the world becomes so complete that it must include a model of itself" (ibid., p. 63).

The deterministic scope of this hypothesis is clearly very strong, and is also challenged in the biological sciences on the basis of equally biological arguments (Sect. 3.5). Moreover, in the second part of his work, Dawkins succeeds in identifying a way out of such a deterministic hypothesis by contrasting the cultural element, the *meme*, with the egoism of genes. With the development of brain and nervous system, humans (and they alone, or so it seems, at least for now) have managed to emancipate themselves from their genome, taking an increasingly greater part in the "policing" of their individual organism. The human species would therefore have found in the meme a new replicating element, based on cultural imitation, which has now taken the place of the gene (Dawkins 1976, p. 214). The proof would lie in the human capacity for a genuine and disinterested altruism, resulting in "the power to defy the selfish genes of our birth and, if necessary, the selfish memes of our indoctrination." Human beings "alone on earth, can rebel against the tyranny of the selfish replicators" (ibid., p. 215).

The concept of the meme had been anticipated, but not fully developed, by Cavalli-Sforza (1971) when he put forward the hypothesis that, in social human evolution, an important mutation has its equivalent in a new disruptive idea. Memes would be replicators in their own right, following an autonomous evolutionary dynamic that is studied by *memetics* (Blackmore 1999, p. 24). This could also be the way morality evolves (Friedman and Sinervo 2016, p. 352ff.). I believe the meme concept had been anticipated in sociology by Gallino (1982, p. 103) with his hypothesis of the replication of cultural and experiential structures. The adoption of children would be an acknowledgment of the memetic need to have descendants for one's own cultural structures.

The idea of the meme is rejected by ultra- and anti-Darwinians alike, precisely because it represents a synthesis, conceding the weight of the cultural factor, without at the same time renouncing the idea that humans are fully biological beings like any other living being (Runciman 2009, p. 53). It is also interesting to note that, though methodologically correct, the idea of the meme has in a sense made researchers lazy, because it tends to be interpreted in a way that is perhaps too obvious and simplistic, that is, as a pure and plain replicator of concepts at the social level (Hodgson and Knudsen 2010, p. 227). Even if in this case there is no declared intent of encroaching on sociology, one cannot but wonder what the reason is for the "horror with which contemporary human and social sciences have looked at naturalistic attempts to interpret political and social facts" (Pennisi 2014, p. 25). In a way, this "biophobia" recalls the reaction that Victorian religious circles had to the Darwinian theory of evolution and to the discovery that we descend from primates. Freese (2008, p. 27) relates what historians of the time attributed to the English nobility:

"Let's hope that [the theory of evolution] is not true, but if it is true, let's hope that it does not become widely known."

Understandably, the mystifying operation of likening genetics and contemporary neuroscience to phrenology or Nazi eugenics and their unacceptable and even evil political outcomes is considered scientifically unsatisfactory (Pennisi 2014, p. 28). The conceptual mistake that traced phenotypic differences between populations and ethnic groups to a pseudo-scientific concept of race (Rose 2007, p. 54ff.) grew out of a genetic theory which had not yet discovered the mechanisms of DNA, and which instead looked primarily at statistical occurrences in morphological similarities. But even if this error had not been made, science could and should have not set itself the objective of "improving" the human species through the culling or forced sterilization of individuals.

Once the dust of the scuffle over sociobiology settled, the idea that it is unlikely that our genome should have no effect on behavior, now more neutrally codified in behavioral genetics, gradually spilled over into the new disciplinary container of evolutionary psychology (Barkow et al. 1992; Laland and Brown 2011, p. 105ff.). Evolutionary psychology and sociobiology diverge in virtue of the different roles they assign to human behavior. For evolutionary psychology, behavior is relevant as the outcome of psychological mechanisms. Sociobiology, on the other hand, limits itself to incorporating evolutionary cues within a sociological theory (Dyevre 2014, p. 368).

Understanding behavior in adaptive or maladaptive terms provides several keys to explaining social behavior. Steven Pinker (2002, p. 319) sets out the three basic laws of behavioral heritability: any behavioral trait is inheritable; the effects of family upbringing are inferior to those that can be related to our genetic behavioral traits; a large portion of the variations in complex behavioral traits cannot be explained in terms of genes or family upbringing. It is clear that, in this way, even sociology cannot have any place for the idea that our behavioral attitudes at birth are a "blank slate." Inheritability concerns overall traits of general relevance, such as intelligence (understood as cognitive abilities measurable through certain tests), as well as behavioral traits of more limited effect, such as dependence on nicotine or alcohol, or a predisposition to divorce (Pinker 2002, p. 321; Lucchini et al. 2013).

On the subject of social phenomena of addiction, it has been hypothesized that in many species evolution induces the development of innate mechanisms of comfort and regulation, from which attachment to others descends, with mechanisms that are similar to that of drug dependence. When the tranquilizing effect of social contact runs out, the brain produces symptoms of abstinence in the form of anxiety and dysphoria. Moderate amounts of stress and anxiety are usually coped with autonomously, with the help of cognitive structures such as memory or cultural systems such as technology or information, and—we could also add by analogy to Frank's (1949) insights—with the help of the law. These substitutes for parental care make their way into the social dynamics because they diminish the importance and influence that physiological mechanisms have on social life (Smith and Stevens 1998, pp. 199, 205).

However, we shouldn't draw hasty conclusions. The apparent simplicity of these behavioral laws can only be understood by introducing a concept that is rarely used in the social sciences, namely, *variability*. Behavioral traits and their heritability are the object of studies designed to better control all the intervening variables affecting subjects who share the same genes (twins) or who share the same family heritage (biological or adopted children and siblings): these studies therefore help to explain what makes individuals different without explaining what they have in common; in other words, they cannot explain what the underlying universal human nature consists of: a 50% variability in some trait does not mean that the behavior is half-inherited; it only means that 50% of the differences between individuals is inherited.

In the criminal field, Raine (2014, p. 44) reports studies on twins to argue that genes may play a role in triggering, but not determining, violent criminal behavior, hypothesizing that while half of this variation would be biological-individual, the remaining share could be divided into a 33% slice deriving from the family environment and a 22% slice deriving from the external social environment. According to Raine, the influence of maternal alcoholism on the predisposition to aggression (ibid., p. 204) is now clearly demonstrated, beyond the subsequent influence that the family environment may have, the importance of which is not necessarily decisive. This finding is based on studies of twins. Although a relevant caveat on twin studies is raised by Vineis (2006, p. 5), who notes that 30% of identical (homozygotic) twins who did not have the placenta in common were prone to a higher level of differentiation, a study on Mauritius that Raine (2014, p. 207) carried out by controlling for twelve intervening social factors seems to confirm a relation between malnutrition (which does not necessarily coincide with undernutrition, but refers to the absence of particular nutrients in the diet) and aggressive behavior.

The second issue is that if the subjects being analyzed differ by macro-variables that place them in different social groups (according to income or education, for example), only the variability internal to each given social group can be explained, not that between different groups. So here, too, we can see that that the analysis of biological factors provides a useful method for understanding micro-variation, but in circumscribed and homogeneous social contexts, and without macro-implications. This does not imply that biological variables have no effect on relevant variables in the social and legal behavior of the individual in the medium range (aggressiveness, drug dependence, affection toward children): it simply means that these variables are not enough to sociologically establish universal laws. Finally, these methodologies only show that some behavioral traits are related to genes, and not that they are the direct and necessary consequence of a genetic cause-effect relationship.

This type of genetic-behavioral analysis distinguishes two very different components, both of which can be characterized as environmental: (1) the shared environment, which consists of the experiences we shared with our siblings (through our parents' childrearing style) and the climate we experienced at home or in the neighborhood; and (2) the non-shared environment, which consists of what exclusively concerned us and not our siblings (when our parents treat us differently from our siblings), and which may also include, for example, a particular disease that has affected only one child. The effects of the shared environment are usually marginal, and in any case not statistically significant, when used to explain differences between siblings (Pinker 2002).

Yet this seems to mark a step forward. We know that some significant behavioral alterations are directly caused by tumors or injuries located in the brain (Damasio 1994, p. 3), even though the strength and direction of such alterations is not always clear. The reference is to the celebrated medical case of Phineas Gage (Pinker 2002, p. 52), a railroad worker who in an accident in 1848 suffered significant damage to the prefrontal cerebral cortex: without apparent functional impairments other than physical ones directly caused by his wound, and therefore retaining all his cognitive faculties and personal memory, he suffered a permanent alteration of personality that reduced him to an asocial subject, unable to feel empathy or to relate appropriately to others (or to resort to hypocrisy when necessary).

In other neurological cases involving the same confined lesion of the brain, the effects were not always the same. In one case, the victim developed restlessness, impulsiveness, and an inability to hold his attention: this made him clearly dysfunctional for complex work tasks, but because he was part of a family and a social network, the condition did not develop into a social disability. In a third case, involving a subject who had suffered the same anatomical damage through a suicide attempt induced by depression and an inability to relate to others, the effect was almost the opposite, making the patient extremely docile and serene (Raine 2014, pp. 133–156).

Our awareness of the biological origin of behavioral traits can extend to a wider area of variability than that of pathological accident. The environmental variable, understood as a family environment, therefore seems largely discredited as a determining factor of social behavior: if we exclude the incidence of unique and accidental events, social behavior is scarcely influenced by our parents' gender, profession, level of education, or marriage status (conventional or gay marriage or single parent) (Pinker 2002, p. 330).

In the argumentation of the social sciences, the most frequent reasoning is of the monovariate kind (Boudon 2009, p. 19), but reality is more like a multivariate continuum of gradations (Vineis 2006, p. 94). The attempt to integrate nature and culture through a biosocial synthesis has met with opposition from an unprecedented alliance of sociological idealism and conservative political ideology (Pinker 2002, p. 337). Sociological idealism correctly denies that behavior is determined by genetic inheritance, but it hits a dead end when it also denies that genetic inheritance can have any influence on social behavior—a denial owed to a fundamental misunderstanding of the biological mechanisms of variability and heritability. This denial is for moral reasons, mistaking the application of behavioral genetics for eugenics, and ignoring that the latter was also instrumentally employed by socialist progressivism (Sidoti and Gammone 2010), as well as by Scandinavian social democracies. This kind of explanation also undermines the importance of accidental events, which, however small, are capable of influencing social behavior in the long run. This is not to surrender to the hard sciences. On the contrary, modern evolutionism acknowledges its debt to the observations the social sciences have made about the behavior of human aggregates (Dunbar 2009, p. 245). The debate on sociobiology

(Ruse 1979; Segerstråle 2000) is echoed in one of the paradigms that could work as a bridge between natural and cultural theories of behavior, that is, *epigenetics* (Goldberg et al. 2007; Pennisi 2008; Goldman 2012, p. 207ff.; Rusconi et al. 2016; Wastell and White 2017; Heil et al. 2017).

Epigenetics (we could rename it "socio-genetic," for it integrates the two perspectives) explains genome modifications that do not involve a DNA mutation: they simply alter the timing and quantity of what is produced by genes at key moments (Rothstein et al. 2009, p. 5). Epigenetics attributes a central role to the transcriptome, that is, to all of the genome expressions that are transcribed as a result of the effect of environmental variables and factors (Robinson et al. 2005, p. 258). This would appear to signal a partial resurgence of Lamarckism, which in this way reveals itself to be compatible with Darwinism, because the latter would instead explain the evolution that occurs through DNA mutations. According to Lumsden and Wilson (1981, p. 2), epigenetic rules feed on information from the physical and cultural environment to create cognitive patterns that form the raw support of thought and decision-making. By further testing and strengthening the genetic hypothesis at the level of human societies, and by showing that the relation between genotype and phenotype is plastic and almost bidirectional, this research would demonstrate that socially induced changes in gene expression constitute a new kind of "social phenotype" (Robinson et al. 2005, p. 268), thus contributing to break the stalemate between the biological and the social (Schnittker 2008, p. 255).

While the existence of epigenetics in animals is now widely demonstrated, we still do not know with certainty in what ways it is expressed or what its intensity is in the specific tissues involved (Miller 2010, p. 27). In mammals, on which the epigenetic hypothesis has seen some initial significant confirmations, the epigenetic relationship is determined by the evolutionary drives that lead descendants to respond to variations in parenting behavior "as a forecast of the environmental conditions they will ultimately face once they become independent of the parent" (Weaver et al. 2004, p. 852). Most experimental confirmations of the effect of exposure to environmental factors come from rodents, whose exposure to particular substances has influenced epigenetic patterns in the development of young subjects with an impact on physiological functions, behavioral expressions, fertility, or cancer propensity, and metabolic diseases once they become adults (Landecker and Panofsky 2013, p. 341).

Foundational research in epigenetics has found that phenotypic predisposition to stress in adult mice is related to exposure to maternal care at a young age. Mothers who spent more time licking and caring for their pups wound up with offspring who were less responsive to stress and who, in turn, would spend more time on their descendants. This effect was confirmed in *cross-fostering*, where mothers with divergent behaviors obtained analogous phenotypes in both the biological lineage and the adopted one (Weaver et al. 2004). In a study conducted on groups of rhesus macaques, where the social rank of individuals was experimentally manipulated, the researchers concluded that "sensitivity to the social environment is reflected in changes in gene expression in the immune system, supporting an increasingly widely recognized link between neural, endocrine, and immune function" (Tung et al. 2012, p. 6494). Being arbitrarily assigned to a social role determined positive or negative changes in the factors of inflammation, which were reflected in individual wellbeing because they activated or deactivated certain genes. In this case, the researchers identified among macaques a causal chain that started from their social rank of dominance or subjection and was reflected in their physiology.

In nonhuman animals we speak on a physiological level of "phenotypic polymorphism," a phenomenon exemplified by social insects. In fact, the most extraordinary epigenetic illustration is that of the social and biological roles that insects take on in a colony. The queen bee is not genetically different at birth from all other workers. It is the nourishment given to the future queen, still in the larval state, that determines the physiological transformations that will make it the only fertile individual in the colony.

It is also true that, for the time being, epigenetic mechanisms in human societies have been reported only in observational or correlational studies (Kaati et al. 2002). This has prevented specific stress factors or time constraints from being controlled, but despite isolated objections (Jörn and Hümpel 2017, pp. 12, 24–26), the accumulated evidence is already quite precise in associating environmental factors, brain functions, and human social behaviors (Sandi and Haller 2015, p. 290; Weaver et al. 2004). Several cohort studies have now revealed long-term consequences for populations affected by catastrophic or war events. A study of 300,000 individuals born before, during, and after a severe eight-month famine in the Netherlands under the German occupation of World War II revealed a noticeably higher incidence of obesity for those who in this period were in the early two quarters of *in utero* fetal development (Ravelli et al. 1976; Roseboom et al. 2000; Veenendaal et al. 2013).

A study on the descendants of Holocaust victims (Yehuda et al. 2016) seems to provide further confirmation that the epigenetic hypothesis is plausible for humans, too. The study offered a demonstration of the way in which preconception trauma suffered by parents is transmitted to offspring, with epigenetic changes in both generations. By way of correlation, several other studies seem to confirm that links among socioeconomic status, stress factors, and health status are transmitted intergenerationally (Baum et al. 1999; Wilkinson and Marmot 2003; Bradley-Springer 2012).

An illuminating metaphor illustrating the relation between genetic heritage and social action is that of a musician who interprets a jazz score freely and with variations (Vineis 2006, pp. 9–10). This is neither about pure improvisation nor about the mechanical performance of the script but is something in between. Epigenetics is potentially a very important step in overcoming the dualism between biological determinism and sociological idealism: a move from nature *versus* nurture to nature *through* nurture (Ridley 2003, pp. 3–4).

A series of behavioral attitudes would crystallize in our biological heritage and would become susceptible to transmission depending on the environment. In this way, attitudes and behaviors acquired through environmental pressures could become hereditary and would be conveyed to subsequent generations. The implications that a disadvantaged social condition reflects on behavioral patterns are clear. One of the behavioral factors behind violent criminal activity, namely, emotional pressure and tension (chronic psychosocial stress), is obviously a feature of the lower socioeconomic strata of the population, this owing to a range of factors: financial insecurity; material deprivation; a poor diet; a lack of health care; exposure to occupational, environmental, and sexual risks; low perceived social status; and low self-esteem (Baum et al. 1999, p. 133). Another link that had previously been noted was between the pain of social exclusion and a lower threshold for physical pain. Studies in environmental epigenetics have also found a correlation between violent behavior and lead intoxication (Nevin 2000, 2007).

Numerous examples illustrate the complexity of the factors behind social behaviors. At least three possible approaches to this problem have been outlined (Pescosolido et al. 2008, pp. 172–173): genetic factors shape social processes hereditarily, either by influencing educational results and the acquisition of human capital or by creating a social safety network through social institutions such as marriage (*social mediators option*); social factors trigger or suppress the expression of certain genes that are involved in particular pathologies, such as alcoholism (*social trigger/suppressor option*); genes attenuate or strengthen the effects of stressors and negative existential events (*genetic attenuation/exacerbation option*).

The resources that individuals can employ from their social network of family, friendships, and associations, and even geographic relationships, lie at the heart of the analysis of *social epidemiology*, and they highlight how social support is useful in coping with social or biological difficulties (Pescosolido et al. 2008, p. 176). For instance, the noticeable increase in obesity rates experienced in Western countries cannot be explained by variation in the genome, but by a change in living habits. Yet it has been shown that obesity then becomes hereditary, and that individual genetic characteristics can moderate the effects that an overconsumption of calories has on weight (Freese 2008, p. 25).

It has been shown that in some individuals with a particular genetic trait, the link between existential experiences and depression, a link previously understood to be indefectible, is in fact practically absent (Shanahan et al. 2008, p. 261). The study of some hormonal biochemical aspects, such as the effects of testosterone on behaviors, has led to hypotheses about their social impact. A high level of individual testosterone has been associated with risk-taking, dominant behavior, aggression, and violence, and it can therefore easily be linked to juvenile delinquency and criminal behavior in adults (Nielsen 1994, p. 288). In men, it has been associated with anti-family outcomes, and in particular with fewer prospects of marriage, a lower marital satisfaction and number of children, and higher rates of infidelity, spouse abuse, and divorce (Freese et al. 2003, p. 245). Testosterone, after all, fluctuates in response to social stimuli, thereby setting in motion a circle of interaction whose origin is difficult to understand, and which can easily be compared to concepts in sociological theory, such as that of "emotional energy," that describe a psycho-physiological state (Collins 1988).

Next to, or instead of, the genome, the study of social behavior begins to refer to the *transcriptome*, that is, to the whole of the genome's expressions, referring to a particular cell or tissue, as transcribed following the effect of environmental variables and factors. While the genome tends to be fixed and stable, except for mutations, the transcriptome is variable by definition (Robinson et al. 2005, p. 258). The recent discovery of "alternative splicing" has shown how the same DNA sequence can be transcribed in different ways (Vineis 2006, p. 11). The social regulation of gene expression is therefore the interpretive key to the synthesis between the social and the biological. Socially induced changes in gene expression would constitute a new kind of "social phenotype" (Robinson et al. 2005, p. 268).

Causation models based on the epigenetic hypothesis are largely relevant for at least two aspects of social-science research: (*i*) class mechanisms and structures and (*ii*) socioeconomic status with related health effects. Epigenetics is a technical innovation not only as a research tool but also in research heuristics. Thanks to this change in perspective, social science may establish a productive relationship with the life sciences by incorporating epigenetic assumptions, just as it has been able to incorporate the physiology of emotions without having to master molecular biology (Landecker and Panofsky 2013, p. 346). To put it simply, "Genomic causation is not in competition with social conditions, but a product of them" (Freese 2008, p. 28). And sociologists are fully equipped to explain how environments can accentuate individual differences in social results.

The epigenetic hypothesis has not been definitively proved in humans, nor can it be expected to in the short term, considering the significant ethical obstacles that systematic experimentation would have to face. As will be explained later, the risk of a prejudicial attitude and persistence of "biophobia" in the social sciences is that sociology should become irrelevant in the public debate (Ellis 1996; Machalek and Martin 2004, p. 461; Freese et al. 2003, p. 234). In this regard, Freese et al. (2003, p. 235) point out the risk that unless sociology can break out of its constructivist fortress, it will be implicitly classified as "creationist." In order to understand mechanisms by which a social position influences a psychological state, we need to understand how genes affect our exposure to the environment, and in order to understand how they affect psychological states, we need to pay attention to social processes (Schnittker 2008, p. 255).

As much as sociology may have understood selective mechanisms, it has not adequately highlighted the strategic or intentional aspects of evolutionarily oriented actions. Although it is clearly difficult to make these research methodologies immediately applicable, given the impossibility of using social biomarkers on humans, methylation mechanisms can provide a heuristic of relationships even in the social field. The methylation process governs the transfer of chemical groups and has the consequence that the genome expresses itself with certain characteristics: it is the basis, for example, of the phenomenon where identical (homozygotic) twins, as they age, tend to increasingly to diverge in their physical appearance, albeit almost imperceptibly (Vineis 2006, p. 15).

The picture is obviously much more complex than the simplified relation between nature and the environment on which many of the hypotheses in the social sciences were based, where one of the factors had to be predominant for the sake of simplicity. This should also caution us about the supposed discoveries of genes suggestively associated with certain character traits or behavioral outcomes (the "madness" gene, the "rape gene," the "math gene"): we can appreciate how false and mystifying these theories are. Even when it was claimed that a gene was likely to be linked to a particular behavior, the most correct statement would have been that a gene was identified that coded for an enzyme that happened to play an important role in the long chain of biochemical processes required to produce some element of the proximate underpinnings of that behavior (Alcock 2001, p. 72). Accordingly there would certainly not even be a criminal antisocial gene, but rather "a number of loci of small effect that together influence temperament, motivation, and cognition and which influence the probability that, in certain environmental and biographical conditions, an individual will engage in an antisocial act" (Rose 2007, p. 240).

Although we find it easier to reason in terms of dichotomies, reality is more like a continuum (Vineis 2006, p. 94). Thus it was once thought that we were on solid ground with the idea of an evolutionary mechanism based exclusively on random mutations, subsequently transmitted more or less successfully on the basis of adaptability to the environment, but even this idea had been challenged by Waddington (1942) through the concept of "canalization." The core of Waddington's thesis is as follows: the reactions that occur during the development of organisms subjected to natural selection are generally canalized, that is, they are adjusted in order to bring about the same final result, without regard to minor variations in environmental conditions (ibid., p. 563). The geneticists' partial victory over the naturalists is not in fact final, and we have yet to explain some phenomena that would seem understandable only by reviving the Lamarckian idea that functions create inheritable traits.

The most recent synthesis in evolutionary theories assumes at least four distinct and interacting dimensions in any type of behavioral dynamic (Jablonka and Lamb 2007): (*i*) the purely structural dimension, given by genetics; (*ii*) the functional and structural dimension, given by epigenetics; (*iii*) the behavioral dimension, given by the choice and conditioning of the environment in which one lives; and (*iv*) the symbolic dimension, given by communicating and acting in a specialized and cooperative way in a society. To elaborate on Vineis's musical metaphor, which compares individual behavior to improvising on a jazz score, interpretations of a particular score that are exceptionally well-executed (owing to the musician's skill or to the quality of the instruments used) can now be recorded using technologies that make it possible for them to be transmitted even more successfully than the original score (whose modification could have happened only by chance or by a mistake in the transcription). In this way these interpretations can become prevalent and maintain that status over long periods of time (Jablonka and Lamb 2007, p. 138).

Epigenetics also carries implications for the law. A knowledge of epigenetic mechanisms could lead to a better understanding of the environmental antecedents leading to certain choices or behaviors: it could thus lessen stigmatization, as it would in fact show that there are no fixed behaviors or irreversible trends. But this adds a further level of complexity to individual variability; so, too, this kind of knowledge could expose personal and health information that needs to be protected as a potential source of discrimination (Robienski 2017), and it could let in hidden forms of genetic determinism (Schuol 2017, pp. 49–50).

3.4 Evolutionary Analysis of Legal Attitudes

Epigenetics, sociobiology, and the evolutionary paradigm can be applied to other disciplines (foremost among them evolutionary psychology and evolutionary biology, but also psychology and behavioral economics) to provide sociology with a new biological perspective on social behavior that the social sciences can no longer afford to ignore. The basic insight of modern evolutionism—developed almost simultaneously by Alfred Russel Wallace and Charles Darwin, the former in some minor works, the latter in the much more famous *Origin of Species* (Darwin 1859)—consists in explaining the evolution of living beings by locating its origin in the combined action of random variation and natural selection, where heritable individual traits mutate at random, and those that get selected achieve reproductive success, thereby passing on to the next generation (Stewart-Williams 2010, p. 27).

The first essay in which the evolutionary theory is sketched out in a comprehensive way (*On the Tendency of Varieties to Depart Indefinitely from the Original Type*) was sent to Darwin in 1858 by Wallace. Darwin had been working on the same themes for a decade: he had hesitated to publish his work because he had not yet managed to explain some inconsistencies—including, for example, extreme altruistic behavior, which until the discovery of inclusive fitness (Hamilton 1964a, b) could not be understood according to a purely evolutionary approach—but now he decided to quickly complete and publish *On the Origin of Species*.

In this section, I will focus specifically on law-informed social behavior, that is, on social behaviors based on legal norms. In fact, these social actions are profoundly influenced by our nature as evolutionarily oriented living beings. This aspect was previously argued with regard to psychological research on the sense of justice (Sect. 2.2). Our drives, biases, and instincts oriented to the basic functions of surviving, procuring food, and reproducing cannot be neglected even in this respect. Evolutionary analysis is effective as long as it is able to integrate with biological studies that identify the causal mechanisms underlying human behavior (Carbone and Cahn 2006, p. 291). The decoupling that morality and law have completed to emancipate from primary needs and impulses does not mean that the latter may be ignored, precisely because in this way we can have a better and complete understanding of the former. Any purely normativist attempt in this sense would be destined for disaster: it is unthinkable to structure a social organization that should fail to factor in the characteristics of its members or the "environmental and historical niche" in which it is located (Scherer 1992, p. 10).

A legal system that neglects the "biological foundation of humanity" ends up being disconnected from the subjects within its jurisdiction (Bohannan 1999, p. 11): such a system would become totally or largely ineffective. In establishing a parallel between the biological and the social sciences, Sampson (1983, p. 57) argues that if scientists want to understand how something works, they must also understand how this thing or phenomenon has come to be what it is. The body of discoveries and additions that this approach has brought to law have been framed today in the biological and evolutionary analysis of law (Alexander 1986; Carbone and Cahn 2006;

Elliott 1985, 1999; Gianola 1995; Jones and Goldsmith 2005; Jones 1999; Jones et al. 2001; Rottleuthner 1985). The fact that evolutionary analysis cannot completely replace other methodologies, especially those of legal theory (Carbone and Cahn 2006, p. 279), does not exclude that it provides precious, if not essential, elements for an understanding of the legal phenomenon.

The evolutionary analysis of law (Jones 2006) places the birth and evolution of legal systems among the fundamental social transitions of human history (Hodgson and Knudsen 2010). A codified system of legal rules is in fact different from a customary habit, and it requires a specialized body of officials that can emerge only in social communities which tend to be sedentary and which have an advanced division of labor. This does not exclude the presence of itinerant bodies of justice, but these emanate from an institutionally if not geographically centralized entity. Legal rules are defined evolutionarily as "a secondary set of norms, adjudicating claims and specifying procedures in social situations of conflict or ambiguity" (Masters 1992b, p. 24).

The development of law takes place in societies that are highly collaborative, as *coordination games* or *self-enforcing institutions* (Hodgson and Knudsen 2010, p. 200). However, this does not explain several cases in which the incentive to pay obedience does not seem sufficient to spontaneously determine widespread compliance. Hodgson and Knudsen concede that the explanation of generalized compliance with the legal system (or at least with the majority of its norms) is psychologically inexplicable, because many laws are still obscure and intuitively difficult to understand.

Contrary to the predisposition to directly punish those who violate a rule—a predisposition entrenched in modern political systems to the point of being itself repressed and punished—the instinct to pay obedience to authority is addressed by cues and suggestions of a cultural nature (ibid., pp. 201–202). Pinker points out that apparently conformist obedience has a distinct rationality. Several cultural customs are arbitrary in their specific forms but not in their *raison d'être*. As we saw in Sect. 3.1, there is no reason why we have an obligation to drive on the right, but the fact that everyone else does it makes the custom more than sensible (Pinker 2002, p. 70). Thus, the differences among countries with respect to numerous important legal rules would be justified by the need for coordination, and therefore cannot be traced back to cultural origins.

The evolutionary approach may be central for law, considering that in the effort to make law more effective, lawyers and politicians are always trying to understand how and why behaviors change in relation to the law (Dyevre 2014, p. 365). The relevance of the evolutionary hypothesis—on which humans today are what they are because of ecological pressures that have pushed them to adapt—is comprehensible in the law if we assume that even the predisposition to morality and law are capabilities that have been developed evolutionarily and adaptively (Conte 1997, p. 84).

Law is a "consumer of behavioral models" (Jones 1999, pp. 272–273), in the sense that it implicitly and intuitively uses simple theories on the motives of behavior, obtaining them from a psychology adapted to the structures of legal thought.

The concept of the "ethological analysis of law" (Gianola 1995) also refers to understanding the law according to the nature of the behaviors it intends to regulate. Law cannot change the genetic heritage unless, following and developing the epigenetic hypothesis, it is ascertained that a sufficiently strong environmental stimulus comes precisely from the legal environment. The law can lead to a reorientation of behavior, but it can also trigger mechanisms of aggression, retreat, or fixation (ibid., p. 820).

Gianola (1995, p. 819) identifies three typical situations in the relation between innate needs and normative stimuli: the behavior being imposed (*i*) conforms to innate behavior; (*ii*) differs from innate behavior but allows the satisfaction of the underlying need; or (*iii*) differs radically from innate behavior and gets in the way of satisfying the underlying need. To contextualize some examples of these assertions at a testable level, we can think of (*i*) the duty to care for and protect children, and the correlative right of children (a hypothesis under the first type of situation); (*ii*) the prohibition on polygamy, accompanied by the introduction of alternatives such as adoption, heterologous fertilization or foster care to protect the innate reproductive interests of individuals (under the second type); and (*iii*) the prohibition on making sexual advances in contexts in which this can create a psychological pressure that violates individual integrity, because the person making the advances enjoys a position of preeminence over the person on the receiving end (under the third type): in this last case, the innate behavior of sexually predatory subjects is discouraged and repressed as damaging to a right.

Concepts such as profit-making, libido, and diligence presuppose motives of conduct which evidently have an evolutionary root, and which the jurist has hidden and perhaps still hides behind the screen of morals. Although Beckstrom (1989, p. 57) is right to say that we cannot expect a direct genetic explanation of each specific behavior, the biological theory would still enable us to better understand drives and build a more robust model of the human action (Jones 1999, p. 270). In this sense, law can also be framed as a cultural universal by which to control emotions (Murdock 1945, p. 124), because it makes it possible to rationally calculate the consequences of acts carried out under the impetus of passions, and it tames the excessive consequences of such passions (Masters 1992a, p. 88).

Under this reinterpretation, there is a strong case that Dyevre (2014) makes in identifying a link between psychological evolutionism and that strain of legal realism which conceives the study of the law as the formulation of predictions about what the decision of the courts will be. Dyevre formulates an interesting hypothesis, suggesting that judicial activism and protagonism of the most prominent judges in American history were in part the result of their "libido." Noting that there is always something sensual in pure altruism, and how several of these great judges also had gained notoriety as womanizers, pure and disinterested selfishness might have been in reality an instrument of sexual conquest (Dyevre 2014, p. 379). There is no denying that this reinterpretation of judicial history may give in to the easy appeal of ad hoc explanations, but that should not be a reason to altogether discard the hypothesis that the most seemingly disinterested altruism, especially if it does not lead to physically dangerous consequences, can often conceal a hidden goal of some other kind. On the other hand, even a display of status (nothing being more alien to altruism) is functional to sexual attractiveness, and so the symmetrically converse hypothesis would in some way also be sustainable. However, this is not to exclude a solution that reconciles both, arguing that judges who embrace humanitarian judicial positions, when the historical timing is right, will achieve greater status and consequently greater sexual attractiveness.

In order to have a more complete picture of the evolutionary niche of human societies at the time when they began to develop a legal system, the evolutionary analysis of law makes use of concepts such as *time-shifted rationality* (Jones 2006, p. 71) and the *law of law's leverage* (Jones et al. 2001, pp. 1144–1145). *Time-shifted rationality* consists in attributing importance to traits which are strictly dependent on evolutionary processes: in some situations they afforded an advantage in our ancestral past, but in the current social environment they also lead to maladaptive and irrational behavior, and they ultimately prove to be disadvantageous (Jones et al. 2001, p. 1172). Examples are aggression and suspiciousness, which in the past may have provided a survival advantage in an environment full of physical hazards, but which, today, could give anyone a bad reputation.

The *law of law's leverage* states instead that "the magnitude of legal intervention necessary to reduce or to increase the incidence of any human behavior will correlate positively or negatively, respectively, with the extent to which a behavior-biasing, information-processing predisposition underlying that behavior (a) increased the inclusive fitness of those bearing the predisposition, on average, more than it decreased it, across all those bearing the predisposition, in the environment in which it evolved and (b) increased the inclusive fitness of those bearing the predisposition more, on average, than did any other alternative predisposition that happened to appear in the environment during the same period" (Jones et al. 2001, p. 1193).

This complex proposition encapsulates a rather simple concept: in some situations we are not able to rationally calculate the consequences of the violation of a norm (Becker 1968). The contribution of evolutionary and behavioral biology consists in reframing emotions, preferences, and other behavioral and decisional mechanisms as evolutionary adaptations (or as subproducts of evolution) rather than as the result of irrationality. The evolutionary adaptations of the past have increased our chances of existential success. This is a different type of rationality from a different era; it is sometimes a long-term type of rationality, even if it may not seem to make much sense from the standpoint of its short- to medium-term utility.

Because of the way in which evolution has shaped our cognitive mechanisms, Jones continues, the legal leverage exerts its effects more fully for those legal rules that directly concern freedom of expression, sentimental relationships, sexuality, privacy, equity, aggression and violence, empathy, status, risk assessment and assumptions, ownership and territory, accumulation of resources, and cooperation and altruism (Jones et al. 2001, p. 1193). It seems now quite clear that "the effectiveness of law will be proportional to the degree to which the function of a particular law complements the function of the behavior that the law intends to regulate" (Gruter 1992, p. 102).

The sociological analysis of behavior applied to law may make it possible to formulate hypotheses that anticipate differences in sensitivity to regulatory incentives or disincentives. Under what conditions will a harsher sanction reduce a given behavior, and to what extent? Why are some acts not discouraged even by the most severe sanctions? In this regard, the evolutionary hypothesis is that the response to immediate existential dangers or opportunities is accomplished through instinctive acts, which do not consider the medium- to long-term perspective. When in times past we would expose ourselves to the risk of committing a violent act, we did not take into account the long-term consequences we are now subject to under the legal system. We have always been aware of this: morality and law have arisen to bend and govern instinct. But we had no clear perception of the biological and evolutionary antecedents of our moral and juridical behavior. Gommer seems to be convinced of this idea in his analysis, according to which "legal systems are the precipitation of proto-morals [that] evolved over millions of years by natural selection" (Gommer 2012, p. 74).

From a statistical point of view, most homicides arise out of seemingly trivial arguments that escalate into violent acts (Jones et al. 2001, p. 1196), and are committed by young men. The most violent acts are therefore in a certain way attributable to subjects who are more sensitive to a perceived threat to their status while this status is under construction. In certain contexts, building a reputation as an aggressive subject, even when risking personal safety, gives enormous advantages (Raine 2014, p. 15). In this sense we can perfectly understand the increasingly frequent recourse to reputational sanctions, even in criminal law, rather than simply inflicting monetary penalties or sentencing offenders to prison.

Among the Yanonamo population, men who have committed murder (they are known as *unokai*) have 1.63 wives and 4.91 children on average, compared to 0.63 wives and 1.59 children for those who have never committed murder (Raine 2014, p. 21). In our ancestral past, aggression and violence have long been an individual character conducive to survival and propagation. Aggressiveness and instinctive aggression toward one's peers has therefore remained widespread, and in certain contexts also advantageous. Predatory killing activates ancestral neural mechanisms of gratification (Wrangham 2006, pp. 56–57): in the accounts of those who have found themselves killing for the first time, it is common to hear that fear and tension turned into a primordial and inexplicable feeling of satisfaction.

Raine (2014) himself highlights a large number of situations in which bio-environmental factors have been found to clearly correlate with the trend in crime rates. Among these, a disconcertingly near-perfect correspondence has been discovered, in specifically identified urban contexts, between the trend in crime rates and the level of lead pollution. Lead pollution reached its peak between the late 1970s and the early 1980s, and then declined with the introduction of environmental regulations. Nevin (2000, 2007) speculates that there is a clear and direct correlation with the rate of violent crimes.

In the areas that are analyzed, reported acts of violence follow a trend that is clearly analogous to the level of lead in the atmosphere. It would of course be possible to advance purely contextual explanations of the phenomenon, assuming that the level of lead is higher in those industrial or postindustrial contexts where, for social reasons, the anomalous and dehumanizing pressures of the capitalist system are also stronger, but Nevin controlled the relation by also using variables relating to social class and family environment, and this would seem to exclude a purely social hypothesis in this case.

The only point where we can have some degree of certainty is that the way in which social and biological factors interact in affecting the likelihood of specific behaviors is so complex that it would be wholly unwise to exclude any one of these elements altogether (Goldman 2012, p. 18ff.). A controversy developed when the MAOA enzyme (Rose 2007, pp. 233, 243) was identified, for the claim was that its variants could determine the aggressive tendencies of a given biological population. This would have seemed to explain, for example, the relatively high Maori crime rate in New Zealand, until it was pointed out that, in ethnic Chinese populations, the gene associated with a dangerous variant of MAOA was even more widespread than among the Maori, yet the rate of homicides remained lower (Raine 2014, p. 55; Vezzoni 2010, p. 71).

As for violent behavior, the simplified models that the social and biological sciences have so far used—simply focusing on the prevalence of a single social, family, or genetic factor—are inadequate, and it will therefore be necessary to resort to models of biosocial interaction (Rabinow 1996, p. 91ff.). Goldman claims, for example, that murder is highly multifactorial, and even the most "lethal" of genetic predispositions do not give us any particularly predictive basis (Goldman 2012, p. 37). The combination of social and environmental factors appears to triple the frequency of antisocial behaviors (Raine 2014, pp. 246–247), and the insistence on a one-factor explanation will lead to unreliable predictions.

In sociological paradigms, the preponderant social factor in explaining obedience to norms is often taken to be that of power or dominance relationships. The motivations to obey norms are grouped by Gruter under three categories: (*i*) the fear of punishment (or, symmetrically, the desire for a reward); (*ii*) group conformism, motivated by both peer pressure and the influence of hierarchical organization; and (*iii*) moral motivations, such as a sense of duty or devotion to an ideal informed by a personal sense of justice. Only the first two categories are "rational," because they can be subjected to cost-benefit analysis (Gruter 1979, p. 44). But domination does not interfere with the development of rules based on reciprocity. In predatory animals like us, the capture of large preys requires a cooperative effort based on reciprocity rules. These rules prevailed even over relations of force and domination. In chimpanzees, for example, it has been observed that the strongest individual never exerts his superior physical power to force a weaker individual to give up a portion of the prey that was hunted cooperatively.

Among baboons, males of higher rank respect the fidelity of females to males of lower rank if the relationship has already had a certain duration—evidence that these higher-ranked males want to avoid destructive clashes of jealousy (Gruter

1979, p. 47). The relational hypothesis mentioned in Sect. 2.5, stating that conflict cannot be completely suppressed, constitutes a sufficient and broad theoretical framework by which to integrate the individualistic competitive perspective and the communalistic cooperative one, if this dialectic comparison is meant as an instrument by which to adjust for mutual social expectations (de Waal 2000, p. 29). This model, derived from the observation of other primates, seems applicable to human social behavior.

Hirshleifer (1980, p. 655) traces a natural history of rights in which he identifies three sociobiological rules that coexist in all communities and shape social and legal rules: the golden rule of "communal sharing," the silver rule of "private rights," and the iron rule of "domination." The establishment of these rules would have favored societies that were more likely to adopt a legal system so as to manage and regulate internal conflicts. These ethological observations can also be applied to human societies, but the paradox of cooperation has not yet found complete acceptance in the social sciences. The very idea of altruism is viewed with skepticism, both in sociology (Bykov 2017) and in the practice and theory of law. Carrie Menkel-Meadow (1992) remarks that in modern adversarial systems the "suppression of altruism" is a *fait accompli*. In current litigation practice, the parties are all but bound by a deontological duty of "mistrust."

Gruter concludes that, in some situations, the rules of reciprocity are more important than power, and she formulates a broader hypothesis according to which the ability to formulate words and develop a language could be the basis of the "human propensity to break down automatically into rules all visual and auditory impressions, leading to a subconscious awareness of rules and, finally, to legal behavior" (Gruter 1979, p. 49). Physical elimination and ostracism by the group—a rather frequent ethological behavior in social species—are indeed one of the first legal sanctions, and in many jurisdictions they still persist as "exclusionary sanctions" (Ferrari 1997, p. 199), such as proscription, exile, banishment, and excommunication (Zippelius 1986, pp. 160, 165).

Yet we can scarcely find among jurists, and even less among sociologists, an explicit sociobiological or ethological approach to the study of law. One of the most interesting modern attempts among jurists is Beckstrom's (1985, 1987, 1989), whose works largely revolve around the implications that sociobiology and ethology have when it comes to interpreting various legal arrangements in North American law. Today, the interdisciplinary study of the relation between law and behavior seems to be the province of evolutionary psychology and behavioral economics, with their research on the law, but the object is substantially the same. According to Beckstrom (1985, p. 3), the sociobiology of law is mainly interested in studying the law's impact on relationships of natural solidarity, such as the provisions regarding wills and probate, the custody of minors, or compensation for injury deriving from the suffering of a close relative.

Lawmakers historically ground their decisions on intuition or common sense. Lawmakers and the courts are currently dealing with problems that are solved "by making guesses as to human aid-giving tendencies" (Beckstrom 1985, p. 147). An interesting example of "familism" accepted in the law is that of the provisions under which, in various ways, one is prevented from testifying *in favor* of a close relative (Slesinger and Hutchins 1929) or is permitted to refuse to testify *against* a close relative (Article 199 of the Italian Code of Criminal Procedure). Sociobiology applied to the law may primarily serve as a support science in the service of improving and calibrating rules (Beckstrom 1987, p. 243). Support is given not only when it comes to guessing the motivations behind a behavior to disincentive, but also when it comes to recognizing that one's motivations are at risk of self-deception. In fact, evolutionary processes have favored the tendency to judge or act in one's own interest without being aware of it (Beckstrom 1989, p. 77).

Beckstrom suggests, for example, that if we are to eradicate nepotism we need to recognize that these behaviors have such a profound biological basis as to deceive us and make us unaware of their impact. "A man naturally loves his children better than his nephews, his nephews better than his cousins, his cousins better than strangers, where every thing else is equal" (Hume 1739, p. 251). This admission is the first step for nepotism to be recognized by those who practice it, and for corrective measures to be taken. This also explains the incoercible drive to know who we are closely related to, not only in order to better address our "altruism" (the tendency to cheat is much stronger toward those whom we do not believe to be our kin), but also to avoid incestuous relationships. The biological predisposition against incest is, however, so strong as to have engendered a taboo that has been translated into a cultural or "biocultural" meme (Beckstrom 1989, p. 64), in this case meaning culture as a variable with elastic boundaries, and governed by epigenetic rules.

To illustrate this hypothesis, we can consider two different cases. When individuals grew up together as children in an Israeli kibbutzim, they later became reluctant to get involved in sexual or close emotional relationships with each other. This is a cultural, nongenetic taboo. The same reluctance can be observed under the Chinese arrangement called *shim-pua* (little daughter-in-law), under which a female child is transferred to a family that adopts her as a domestic worker and then marries her with one of its sons. Couples formed through this arrangement were reluctant to establish the relationship and, once married, had a fertility rate lower than one-third, and a higher rate of divorce (Beckstrom 1989, p. 61). The prevalence of the cultural taboo over the genetic one is also known in anthropological literature on matrimony as the Westermarck effect (De Tullio 2001).

Evolutionary and sociobiological paradigms manifest themselves even more clearly in the regulation of conjugal fidelity and parent-child relationships. It was only recently that female adultery ceased to be held to a double standard under the law, and this was not only on account of tradition but also to advance a reproductive interest: in a traditional setting, it was the adulteress's husband who was considered the victim in adultery, and never the adulterer's wife. This was consistent with the biological makeup of the species, where the father tends to cooperate in providing for the family while the children are growing up, and yet, until recently, there was no way to establish paternity with absolute certainty (Nielsen 1994, p. 284). In traditional morality, and consequently in the law, the risk of devoting one's parental investment to a stranger's offspring justified a more severe punishment for the wife. In societies where the incidence of sexual promiscuity is high, men tend to divert their support and attention, at least in part, toward the children of sisters, where the biological bond is certain (Beckstrom 1989, p. 51).

Several observations can also be made about polygamy and monogamy. Although polygamy is anthropologically quite common, the most numerous societies, and hence those that can evolutionarily be considered "winners," are founded on legal principles that protect monogamy, and monogamy mainly protects equality of reproductive opportunity for men. If the most powerful and influential individuals were allowed to establish parenting relationships with a large number of women, a greater number of males would be left without a wife and without offspring. The protection of monogamy can thus be construed as a form of "male solidarity," established in the interests of social peace (Wright 1995, p. 92ff.).

The importance of the "biological investment" is further confirmed by statistics on cases of violence against minors, which show that abuse of children is markedly more frequent among adopted children (Daly and Wilson 1997, p. 118) and, moreover, that infanticide (Jones 2006, p. 70) mainly takes place in the first year of a child's life, when the parental investment in them is still limited (Jones and Goldsmith 2005, p. 433). There are no significant differences in the frequency of violence committed by mothers in comparison with fathers, thereby excluding that a mother's postpartum depression can be a decisive factor (Raine 2014, p. 26). Finally, most infanticides are committed by young mothers, lending credence to the hypothesis that because these women are young, and so have a greater reproductive potential than older women, the act can in this case be seen as more acceptable (Raine 2014, p. 28). As these examples suggest, biology is often determinative when making moral decisions concerning life or death or in acts of altruism.

The biological drives at work in committing acts the law considers to be bad or severely damaging are considerable. This requires a more advanced reflection in terms of lawmaking, too. Two main options can be hypothesized. Since certain acts are for some people more likely, because they are more instinctive, if we were to follow a preventive logic it would be necessary for the law to provide even more severe sanctions for the individuals so predisposed, in such a way as to more effectively counteract the biological drive. This is, for example, what Beckstrom (1985, p. 133) proposes in cases involving adoptive parents committing violence against their adopted children. On a more retributivist and protective logic, the biological predisposition could be considered a mitigating factor, or else as a determining factor in a diminished mental capacity, and could at least contribute to balancing out the aggravating circumstances that generally accompany acts of violence against a vulnerable person in one's care.

Another issue, naturally, is that of succession, both in cases of intestacy and in cases involving testators disposing of their estate by will in countries where freedom of will is not absolute. Here the lawmakers seek to "proportionally" estimate the value of blood relations, and in some jurisdictions the law even provides that the closest relatives, except in special circumstances where they are deemed undeserving, in any event have the right to a portion of the estate. The point made by Beckstrom (1985, p. 58), however, is that lawmakers attempt to construct a legal framework for intestate succession that often reflects past cultural conceptions. Another deeply rooted instinct, equal to the instinct to own property, is to make sure that one's own estate, or a non-negligible part of it, passes to descendants or family members (Stake 2006, p. 193ff.), and this explains the resistance to high estate or inheritance taxes.

Also a sociobiological theme *par excellence* is that of disputes and decisions regarding the custody of children. By virtue of previously mentioned asymmetries, lawmakers have tended to presume that the more attentive parent, and so the one who should get custody, is always the mother. Even here, however, the asymmetry is diminishing as a result of cultural mutations that bring in an analogy to adultery: if adultery is symmetrically evaluated between man and woman, so must issues relating to parenthood and the care of the children. Equal treatment of children in inheritance is a modern trend that has undermined the preference accorded to the firstborn male (Stake 2006, p. 198).

The evolutionist paradigm, especially in the early stages of its taking hold, has been used not only to explain individual social mechanisms but also to describe the evolution of normative systems (Barberis 1998; Elliott 1985, 1999; Gianola 1997; Keller 1919; Salvadori 1904; Zamboni 2010). Salvadori (1904, pp. 12–13), for example, foresaw that evolution would lead to a "perfectly contractual society," in which the individual could be increasingly autonomous and emancipated from external coercion, this on the basis of the harmonious cooperation which was slowly taking hold by adaptation through "the perfecting of the legal conscience" (ibid., p. 32). Salvadori paid tribute to a "philosophy of evolution" that offered a naturalistic perspective by which to explain the advancement of the positivistic paradigm of science, and ultimately of legal scholarship. His tribute is condensed in this passage: "We need to credit the philosophy of evolution with having demonstrated, through a rigorously positive investigation, what the true foundation of legal norms is. With the progress of the natural sciences, and the new physical, biological, psychological, and sociological theories, came a great transformation in humanity's general ideas about the universe; the origins and development of life, thought, and society; and the genesis of the moral sense: this transformation could not fail to affect legal philosophy" (ibid., pp. 8–9).

In spite of the endemic conservatism of lawyers, innovative and original contributions have come from jurists adopting evolutionary theory (Elliott 1985, p. 93). It is widely thought that the sociological approach developed by Roscoe Pound was at least partly influenced by his studies in botany. According to Barberis (1998, p. 45), the evolutionary approach to the study of change in legal systems implies something akin to the Hayekian principles: not specific predictions but hypothetical models of future courses of action. On this basis he concludes that it is not possible to extract any moral or legislative guidance in a broad sense, thus rejecting the constructivist criticism that the use of the methodological tools of the natural sciences would ultimately lead to a neoliberal outcome.

The biological analogy proves to be powerful enough to interpret and also explain the relations that hold *between* legal systems (Elliott 1999, pp. 112–113), even though the daily life of norms unfolds in a *micro*-state (ibid., p. 123). The law becomes a cultural mechanism similar to the meme, subjected as it is to ecological pressures and therefore forced to evolve. Laws are transmitted through a system of replication and copying, and they apparently spread on the basis of their effectiveness in adapting to a particular social system. The original law and its copy thus constitute a "conditional generative mechanism" (Hodgson and Knudsen 2010, p. 205).

Elliott (1985, p. 40ff.) identifies four major strands in the evolutionary theory of law, each based on a different analogy between biological evolution and law. (1) In *social* evolutionary theories, the law evolves with the evolution of society, and as an integral part of it. (2) In *juridical* evolutionary theories, norms evolve and survive by adaptation. (3) *Economic* evolutionary theories postulate that the law evolves on the basis of the choices made in litigation: not the choices that judges make in interpreting the law, but those the litigants themselves make in deciding which problems to solve within the framework of the law itself. (4) Finally, there are *sociobiological* evolutionary theories, even if, on closer inspection, Elliott (1985, p. 40ff.) in this case conflates two distinct problems: (a) the interference of biological mechanisms in legally oriented activity and (b) the evolutionary nature of change in legal systems. It seems to me that even Sacco (2015), in developing his concept of "mute law," in essence adheres to an innatist conception of the sense of the law.

In conclusion, it is possible to argue, taking the perspective of legal philosophy, that law is a substantially evolutionary phenomenon, and that it can be studied with the conceptual categories of biology (Gommer 2012, p. 59), and in particular through the notions of replication, stability, and reciprocity. That certain rules of conduct carry an evolutionary advantage is a hypothesis also advanced by von Hayek (2013, pp. 140, 183). From these premises, however, it cannot be inferred that law is ultimately a mediated emanation of genes, as Gommer (2012, p. 62) seems to assume, simply because it would derive from the evolutionary nature of human proto-morality. For this would indeed be a legal version of social Darwinism (Hoffman 2006, p. 3).

Having defined this framework, we can outline a comprehensive definition of legal rights. Legal rights—in both the active situation of the right *to do* something and the passive situation of the right *to not do* something or *not have something done to us*—arise from a moral heuristic located in our instinctual cognitive system, that is, from a filter of judgment that leads us to classify some actions as opportunistic, egoistic, or manipulative. Legal rights define themselves as such by a process of cultural elaboration in which the cognitive emotions selected on the basis of this moral heuristic are reconstructed by our *rational* cognitive system according to institutional and collective needs. Thus, the feeling that something is a legal right that we can assert for ourselves is filtered and redeveloped through the collective notion of what ought to be considered or promoted as such.

3.5 Criticism of the Biosocial Synthesis

In concluding this chapter on the evolutionary and sociobiological analysis of law and behavior, it is necessary to summarize and outline a response to some of the objections raised against the use of biosocial paradigms (Rose and Rose 2014). It seems to me that the most balanced supporters of the integration between biological and social explanations (Alcock 2001; Pinker 2002; Wright 1995) have been able to find an acceptable balance between environmental and genetic factors, weighing the respective variables according to the relevance they are assumed to have.

However, a large group of scholars in biology who side with constructivist sociology have tried to keep a separation between biological explanation, applicable only to *nonhuman* animals, and social explanation, applicable to *human* animals. Between the 1970s and the 1980s, evolutionary theories that sought to explain social events were enlisted in what Segerstråle (2000) describes as a "battle for science." This is apparently a controversy with political nuances between conservative/innatist positions and progressive/culturalist positions. The underlying interests are different from the accompanying political orientations and cut across the latter, having perhaps more to do with academic status (Segerstråle 1990, p. 273).

As anticipated in Sect. 3.3, one of the best-known episodes, illustrating how purely scientific issues can intersect at several points with the political climate of the period, is the debate that a study group conducted on sociobiology (or rather *against* it) by inviting Noam Chomsky, with Segerstråle participating as an external observer. On that occasion, Chomsky, universally known for his radical and progressive political ideas, refused to speak against the sociobiological hypothesis. In the debate with the biologist Lewontin, the latter rejected the sociobiological position on the basis of an interpretation of human nature closer to the older and "dialectical" Marx, while Chomsky could not adhere to the vision of a socially constructed humanity, and defended the presence of universal and innate traits, subscribing the positions of the younger Marx (Segerstråle 2000, p. 204). Chomsky's political radicalism was motivated by a partly biological view of human nature (ibid., p. 205).

The criticism explaining why it would be misconceived to introduce biological elements into social science has been articulated by theorists of biopolitics, and hence, in the first place, by Foucault. Although Foucault was open-minded in addressing the opposite positions (Chomsky and Foucault 2006), there remains an unbridgeable difference. In Chomsky's innate vision, justice is rooted in fundamental human qualities, while in Foucauldian constructivism it is a social construct (ibid., p. 100). In reality, "biopolitics" is a neutral connotation that can qualify diametrically opposed visions of the world: from the objectivistic and *tanatopolitical* biopolitics of Nazism to the subjectivist, aesthetic-Dionysian biopolitics of the hedonistic Marxism embraced by Marcuse and, eventually, by Foucault himself (Pennisi 2014, p. 27). An equally critical view of the biopolitical stance in contemporary political systems has also been expressed by Rose (2007), but his position is mediated by an undoubted understanding of biological implications and problems.

In the biopolitical vision, the modern cognitive sciences would amount to a biologism impoverished by economism and would be an outgrowth of neoliberalism (Garapon 2012), which is to say that they would be an instrument in the hands of "cognitive capitalism." The association between Darwinian evolutionism and Nazi eugenics is evidently ingrained, but the sociological critique of the evolutionary principle was deeply rooted well before the World War II and its concentration camps (Hodgson and Knudsen 2010, p. 16; Vignera 2010, p. 192).

Some criticisms proceed on a properly methodological basis from the point of view of biology. For example, Symons (1992, p. 146) claims that Darwinian social science (which he labels "Darwinian anthropology") is not really Darwinian in the argument it makes that human behavior continually adapts to maximize reproductive success. It has also tellingly been pointed out that when the concept of evolution is used as a metaphor, it ceases to perform any useful empirical function (Gladue 1989, p. 255).

The critical biology propounded by Maturana and Varela (1980) instead takes the autopoiesis of biological-social systems as its starting point, clearly bringing the Luhmanian concept of "structural coupling" to bear on biology: "we as social beings generate, through our structure-determined properties, our societies as the cultural media that select our individual paths of ontogenic change in a manner that leads each one of us to the structure that makes us generate the particular societies to whom we belong" (ibid., p. xxvii). From this perspective, the authors suggest, the adoption of the Darwinian notion of evolution looks like another attempt to justify certain despicable aspects of human existence, among which "social discrimination, slavery, economical subordination and political submission of the individuals, isolated or collectively, to the design or whim of those who pretend to represent the values contained in those notions" (ibid., p. 117). On the basis of the autopoietic organization of the individual, they claim that biology should not be used to make the individual expendable (ibid., p. 118).

Other criticisms specifically address the evolutionary analysis of law and the application of evolutionary biology to law, dismissing this as a "fad [with] a shelf life at least as short as deconstruction's" (Leiter and Weisberg 2009, p. 34). It is interesting to dwell on the criticism that Weisberger and Leiter direct at the systematic application of the biological perspective to legal behavior, because the authors take up a classic positivist argument also deployed against the sociological analysis of law. The central point is why law should deal with how certain behavioral traits have evolved, and not focus instead on how people behave, here and now, since the "etiology of behavior" remains essentially irrelevant to the issues of legal regulation (ibid., pp. 35–39). Leaving aside the debate in biology on whether or not natural selection is the primary mechanism governing the evolution of living organisms, Leiter and Weisberg doubt that biology can still make a significant contribution to the explanation of behavioral and social phenomena (ibid., p. 60). But the criticism boils down to an epistemological objection to the fecundity of interdisciplinary approaches in general, so much so that in order to strengthen the concept, they argue that even critical legal studies or the economic analysis of law have already entered a phase of decadence (ibid., pp. 33–34).

These kinds of remarks have already been addressed in Sect. 1.5. Beyond the fact that such decadence is anything but proven, it is hardly necessary to point out that although several of the hypotheses of evolutionary biology are currently contested, or have been radically challenged under other epistemological paradigms, it cannot certainly be concluded on that basis that they are irrelevant for any open-ended, accommodating vision of the law. And even if this accommodating vision is not universally shared, a solid rebuttal of it cannot simply rest on its characterization as a "fad." The view that "critical attack by social scientists on sociobiology provides abundant evidence of misquotation and gross misrepresentation" (Hodgson and Knudsen 2010, p. 20) is certainly not isolated. The view that scientific knowledge, in particular, is socially constructed—and therefore part of a struggle for power that is eminently social and not at all biological—appears to be widely shared in certain milieus. And this is definitely not just the isolated position of "radical relativists who believe that all knowledge is socially constructed" (Jones and Goldsmith 2005, p. 422).

However, the Foucauldian biopolitical paradigm and the sociological approaches that follow it are somewhat blind to the important concepts of structural, morphological, and functional constraints. These concepts are the basis of modern biology and constitute "a limit to the anthropocentric omnipotence of human thought and of its infinite creativity" (Pennisi 2014, p. 34). The Standard Social Science Model, as previously defined, has artificially separated the biological from the social, safeguarding the political correctness of scientific inquiry, while often crossing over into Lysenkoism. Paradoxically, it was precisely for this reason that it failed to explain the differences between human societies without resorting to the concept of race. It is now understood that the external morphological differences between ethnic groups are genetically marginal: these differences can be much smaller than those existing among members of the same ethnic group, and this also refutes, even biologically speaking, the very idea of race (Pinker 2002, p. 134). As we have seen, an answer that is already quite complete has come from evolutionary psychology. It is not possible to assume that some individual traits or abilities that develop in the course of individual growth, such as sexual organs or dentition (Tooby and Cosmides 1992), or language (Chomsky 2002), are cultural and not innate just because they are not present at birth.

Sociobiology has also been used as a "subversive" and progressive discipline, precisely because it highlights that organisms do not exist for the benefit of their kin group or their species (Trivers 2002), and that social hierarchies are not a natural necessity for the welfare of the species, thus eliminating an important ideological justification for inequality (Wright 1995, p. 257). For a good part of the twentieth century, eugenics was invoked by progressives and certainly not by conservatives or reactionaries (who claimed it was preferable not to interfere with the natural state), and was universally rejected only when it led to the aberrant practice of forced sterilization (practiced in various Scandinavian countries) and Nazi experiments.

The historical concern about the "terrible truth" of human nature is traced by Pinker (2002, p. 130) to several fears: the recognition of an innate difference between individuals would legitimize discrimination; the discovery of cases of innate immorality would make the hope of improving the human condition empty; the preponderant role of biology would erase the relevance of free will and individual responsibility, and would make life a "baser" exercise deprived of higher moral expectations. Today, at any rate, biological determinism is disavowed by genetics (Alexander 1986): genome mapping has simply confirmed that there are not enough genes to justify the huge diversity in traits in the human species.

We already considered the clash between culturalism and innatism/psychology in Chap. 2, but the issue must now be analyzed in light of the sociobiological hypothesis. I feel confident that sociology should be able to find a synthesis, in the first place by subscribing to this position advanced by Crespi (1989, p. 139): "If it is true that culture forms social actors since birth, and if it is true that cultural forms are sometimes absolutized to the point of being smuggled as nature, it is also true that culture never succeeds in achieving a deterministic force equal to that of instinctual mechanisms in animals, or in incorporating all the different manifestations of lived experience."

Malinowski was extraordinarily prescient when he stated that "bodily needs and environmental influences, and the cultural reactions to them, have thus to be studied side by side" (Malinowski 1939, p. 940), and that "the creation of cultural symbolism and values, cannot be understood without direct reference to individual psychology and physiology" (ibid., p. 956). In another passage from this important work, Malinowski reveals a quasi-epigenetic conception of the relationship between culture and human biological structure: "Culture thus appears first and foremost as a vast instrumental reality—the body of implements and commodities, charters of social organization, ideas and customs, beliefs and values—all of which allow man to satisfy his biological requirements through cooperation and within an environment refashioned and readjusted. The human organism, however, itself becomes modified in the process and readjusted to the type of situation provided by culture. In this sense culture is also a vast conditioning apparatus" (ibid., p. 946).

On the sociological side, Bearman (2008) rationalizes sociology's distrust of the sociobiological hypothesis by reducing this distrust to the sociological stance against (1) a resurgent functionalism, (2) a radical overturning of sociological epistemology, and (3) the resurrection of a eugenic political project. However, these changes would probably not amount to a paradigm shift but would rather entail adding a further level of variables to sociological analysis (Ghezzi 1987, p. 112ff.). From this cognitive perspective, the appeal made to the Darwinian evolutionary paradigm is a guiding model to be understood in a nondogmatic sense. The paradigm is not meant to be used to predict or explain any contingent behavior, but would still be valid as the backbone of a unified evolutionary theoretical framework for the social and behavioral sciences. To so identify a true human nature is not to invite racism, sexism, or aggression but to embrace "a realistic, biologically informed humanism" (Pinker 2002, p. 15). As Cimatti (2013, pp. 15, 21) critically remarks, the strictly anthropocentric vision would lead to the unacceptable conclusion that only humans are able "to distance themselves from their own animality" and "relate to beingness."

It seems appropriate to return once more to the *tabula rasa* fallacy, perhaps the founding cornerstone of the Standard Social Science Model (Marraffa 2011, p. 167). The assumption that our mind is vacuum to be filled is the dream of every dictator (Alcock 2001, p. 153), but this is not the point. The point is that the modern sciences concerned to describe and understand behavior have moved past the dualism between mind and nature (Bateson 1979), and past the simple choice between nature and culture, recognizing that living beings are a mixture of both, in different proportions depending on the context (Greenberg and Cohen 1982, p. 84). Pattaro (2005, p. 365) convincingly argues that "the very distinction between nature and culture is not natural but cultural. Even the notions of nature and culture are cultural." Pinker (2002, p. 67) suggests that culture could ultimately be identified with those neural circuits that deal with learning. In this imaginary dialogue between society and nature, Wentworth and Yardley (1994a, b, p. 26) reconstruct the relationship as follows: "Through the intermediaries of our social psychology and the body, society has told the genes, 'Let go, we are doing fine.' The genes have apparently agreed. Their reply: 'Do what you will, but go forth and form social bonds. That is our parting command.'"

The example of gender differences illustrates an important aspect of this issue. According to the Standard Social Science Model, humans do not have an essential sexual nature, and the only real differences between men and women would lie in their reproductive functions, since all the other gender differences are the product of differentiated social conditioning (Browne 1999, p. 75). However, the biggest meta-analysis of the origins of gender differences tells a different story (Lytton and Romney 1991). The analysis looked at 172 studies involving over 28,000 children of both genders. The result—considering a long series of parameters that included encouragement, warmth, support, severity, discipline, and clarity in communication—was that parents do not treat males and females differently. The only substantial difference this meta-research found was that two-thirds of boys are actively discouraged from playing with dolls, especially by their fathers, out of fear that in this way they could grow up to be gay. Although children who exhibit a preference for girls' toys actually do have a greater likelihood of becoming gay in adulthood, prohibiting the use of such toys does not seem to have any effect on the end result. On the more general problem of the construction of gender identity, Pinker (2002, p. 360) polemically notes a certain hypocrisy: "Those who argue that gender differences are a reversible social construction do not treat them that way in their advice to their daughters, their dealings with the opposite sex, and their unguarded gossip, humor, and reflections on their lives."

Biological and cultural differences are not so easily distinguished. The canons of beauty, for example, attest to a substantial uniformity between cultures and ethnic groups in the criteria of judgment, both for males and females (Alcock 2001, p. 140). In terms of family behavior in humans, both men and women tend toward a monogamous family structure (Stewart-Williams 2010, p. 234) despite frequent deviations into promiscuity, which likewise occur in both sexes, and with differences that can instead be explained in terms of reproductive functionality. The human ability to quickly develop and learn new behaviors has been facilitated by language, and has led to the conclusion that there exist some "social instincts" which would develop under the pressure of evolution over a few generations (Richerson and Boyd 1999, p. 265).

The cultural system can therefore be considered a much faster-paced subspecies than the biological-genetic adaptive system, and is so important as to have become almost preponderant. It has given humans an outsized evolutionary advantage, owing to the ability to respond with a speed never before seen in the animal world. Morality, law, and religion can be considered evolved biological-cultural instincts: rapid and cumulative cultural changes (Runciman 2009, p. 10), a source of progress as well as of problems and obstacles never before seen in the system of the living (anomie, manipulation, wars of aggression), at least not in such massive forms. These problems of transition, according to van Schaik and Kai (2016, p. 75), are dealt with through books of religious prescription such as the Bible. To cite the automotive metaphor of two evolutionary scholars, humans would be "designed for speed, not for comfort" (Richerson and Boyd 2001, p. 451).

In this chapter I have tried to lay the theoretical foundations for some research hypotheses that will be sketched out in the closing part of this book. In what concerns the most controversial points, it seems to me that, while the accusation of determinism is probably reasonable, it would be more difficult to accept that of reductionism. And if this latter objection should turn out to be correct, I might settle for "refined reductionism" (Di Giovine 2009, p. 192). There is a possibility for the behavioral sciences to predict the behavior of individuals, but this will not be possible through exclusive or even prevalent recourse to only one of the cultural or biological dimensions (Goldman 2012; Mesoudi et al. 2006).

From an evolutionary point of view, there is no conceptual leap between humans and monkeys, just as there is no leap between monkeys and pine trees. Anthropoid primates could evolutionally develop the faculty of language, just as humans have done. As far as we know, vegetables are able to communicate through biochemical signals (Yoneya and Takabayashi 2014) and with animals (Leonard and Francis 2017). Bacteria send and receive chemical social information about the presence and status of other bacteria in their social context and act accordingly (Nettle 2009, p. 226). Konrad Lorenz's description of dogs as animals with a conscience get us to wonder if their awareness of our expectations, and of the punishments they receive for disobeying, does not cause even socially simpler animals to experience something akin to the sense of guilt (de Waal 1992, p. 246). Gallino (1987, p. 21) acutely observes how the assumptions behind the separation between theories of human and animal behavior have already been called into question one by one.

It also seems to me that certain bedrock points have been established, such as that natural selection should not be used as a way to assert moral rules or legitimize the *status quo* (Stewart-Williams 2010, p. 232). But it does not thereby follow that natural selection must also be discarded as a general explanatory principle for social phenomena, or as a "research program" (Popper 1978, p. 222). In fact, there is an argument to be made that natural selection can be so used, since biology and the social sciences share the same rules (Rossetti 1982, p. 118). According to Kull (2016, p. 645), the paradigms of semiotic biology could, for example, reestablish the link between functionalism and structuralism, thereby proving suitable for the social sciences and the humanities, as well as for biology.

I fully subscribe to the idea that the evolutionist paradigm, understood in this secular and modern sense, can provide a strong enough basis on which to resolve the dualism between holistic and individualistic approaches and between idiographic and nomothetic explanations. The neo-Darwinian approach is not opposed to the social and cultural determination of behavior: it rather offers itself as a framework for understanding these determinants (Nettle 2009, p. 224) and recognizing that social and cultural factors are biologically mediated (Wright 1995, pp. 348–349). In this regard, the epigenetic hypothesis has provided the missing link, making it possible to appreciate that the path between genotype and phenotype is bidirectional. Lacking a medium-range theory that serves as a link to more precise predictions (Hodgson and Knudsen 2010, p. 231), the sociological application of evolutionism must be able to somehow use auxiliary theories to come down to a lower level of abstraction.

One remaining difficulty, provided that natural selection is operative at the societal level, is to determine whether it has any clear direction: certainly it is difficult to claim that it does in biology. Can we at least accept that we have averted the risk that Durkheim highlighted when he cautioned against a sociology that gets flattened into a sort of "social biology of unconscious and arbitrary phenomena" (Marra 1984, p. 21)? When Peter Singer (2000, p. 6) pointed out the need for a "Darwinian Left," he was urging us to appreciate that human nature works in a modern sense, in that "we are evolved animals, and [...] we bear the evidence of our inheritance." Cooperation studies show that human nature, as well as that of any living being, was originally inclined toward cooperation, and not just toward competition. The most social regulation favors forms of cooperation, discourages deception, and attempts to channel competition toward desirable social ends (Singer 2000, p. 61).

The study of isolated genes in microorganisms is now unveiling revolutionary aspects in the social order of the living. Many genes with social effects have been identified (Foster et al. 2007, p. 79), but the path from genotype to social phenotype is not clear, and it is presumable that the more we try to go into detail, the more we will face greater and greater conceptual and methodological difficulties. I believe it is quite plausible, in this research program, that the sense of justice and the legal sense can be extensively interpreted as a set of adaptive biological-cultural traits. People generally tend toward cooperation, because it is individually and therefore socially useful, and law favors this tendency, but substantial niches of free-riders remain and will remain. These persons eat away at significant chunks of the common welfare, but every measure taken to disadvantage and discourage such behaviors must be carefully weighed, because we would risk inhibiting the naturally cooperative behavior of the bulk of the population. In fact, the current debate on lawmaking policies and instruments is focused on the problem of transitioning from a model of paternalistic regulation to an incentive/disincentive model based on human interests (Baldwin 2014; Thaler and Sunstein 2009). This transition is the result of an evolutionally and biologically informed model of law that recognizes that the best way to direct behavior is to leverage the instincts we have evolved which incline us toward sharing and cooperation.

References

Addario, Nicolò. 1994. Da Parsons oltre Parsons. Una soluzione epistemologia del nesso micro-macro. In *Il rapporto micro-macro nelle teorie sociologiche contemporanee*, ed. Nicolò Addario. Milano: FrancoAngeli.

Alcock, John. 2001. *The Triumph of Sociobiology*. Vol. 93. New York: Oxford University Press.

Alexander, Richard D. 1986. Biology and Law. *Ethology and Sociobiology* 7: 20.

Axelrod, Robert. 1984. *The Evolution of Cooperation*. New York: Harper Collins.

Axelrod, Robert, and William Donald Hamilton. 1981. The Evolution of Cooperation. *Science* 211: 1391.

Baldwin, Robert. 2014. From Regulation to Behaviour Change: Giving Nudge the Third Degree. *The Modern Law Review* 77 (6): 831–857.

Barberis, Mauro. 1998. *L'evoluzione nel diritto*. Torino: Giappichelli.

Barkow, Jerome H., Lela Cosmides, and John Tooby. 1992. *The Adapted Mind. Evolutionary Psychology and the Generation of Culture*. New York, Oxford: Oxford University Press.

Bateson, Gregory. 1979. *Mind and Nature. A Necessary Unity*. New York: E.P. Dutton.

Baum, Andrew, J.P. Garofalo, and Ann Marie Yali. 1999. Socioeconomic Status and Chronic Stress: Does Stress Account for SES Effects on Health? *Annals of the New York Academy of Sciences* 896 (1): 131–144.

Bearman, Peter. 2008. Introduction: Exploring Genetics and Social Structure. *American Journal of Sociology* 114 (S1): v–x.

Becker, Gary S. 1968. Crime and Punishment: An Economic Approach. *Journal of Political Economy* 76 (2): 169–217.

Beckstrom, John H. 1985. *Sociobiology and the Law*. Urbana, Chicago: University of Illinois Press.

———. 1987. The Use of Legal Opinions to Test Sociobiological Theory: Contract Law Regarding Reciprocal Relationships in a Household. *Ethology and Sociobiology* 8: 243.

———. 1989. *Evolutionary Jurisprudence. Prospects and Limitations on the Use of Modern Darwinism Throughout the Legal Process*. Chicago, Urbana, IL: University of Illinois Press.

Bertacchini, Pier Augusto, and Maria Luisa Genta. 1982. Lo studio biologico del comportamento sociale. In *Sociobiologia possibile. neodarwinismo e scienze dell'uomo: la ricerca di un'alternativa al determinismo biologico*, ed. Marco Ingrosso, Sergio Manghi, and Vittorio Parisi. Milano: FrancoAngeli.

Blackmore, Susan. 1999. *The Meme Machine*. Oxford, New York: Oxford University Press.

Bohannan, Paul. 1999. One Anthropologist's Gratitude to the Gruter Institute. In *Law & Evolutionary Biology*, ed. Lawrence A. Frolik, Wolfgang Fikentscher, and Gerti Dieker. Portola Valley: Gruter Institute for Law and Behavioral Research.

Boudon, Raymond. 2009. *Effets pervers et ordre sociale*. Paris: Quadrige - PUF.

Bradley-Springer, Lucy. 2012. The Social Determinants of Health. *Journal of the Association of Nurses in AIDS Care* 23: 181–183.

Brosnan, Sarah F., and Frans B.M. de Waal. 2003. Monkeys Reject Unequal Pay. *Science* 425: 297–299.

Brosnan, Sarah F., Owen D. Jones, Susan P. Lambeth, Mary Catherine Mareno, Amanda S. Richardson, and Steven J. Schapiro. 2007. Endowment Effects in Chimpanzees. *Current Biology* 17 (19): 1704–1707.

Browne, Kingsley R. 1999. Law, Biology, Sex and Politics. In *Law & Evolutionary Biology*, ed. Lawrence A. Frolik, Wolfgang Fikentscher, and Gerti Dieker. Portola Valley: Gruter Institute for Law and Behavioral Research.

Burkart, J.M., O. Allon, F. Amici, C. Fichtel, C. Finkenwirth, A. Heschl, J. Huber, et al. 2014. The Evolutionary Origin of Human Hyper-Cooperation. *Nature Communications* 5: 4747.

Burnstein, Eugene, Christian Crandall, and Shinobu Kitayama. 1994. Some Neo-Darwinian Decision Rules for Altruism: Weighing Cues for Inclusive Fitness as a Function of the Biological Importance of the Decision. *Journal of Personality and Social Psychology* 65: 773–789.

Bykov, Andrey. 2017. Altruism: New Perspectives of Research on a Classical Theme in Sociology of Morality. *Current Sociology* 65 (6): 797–813.

Carbone, June, and Naomi Cahn. 2006. Examining the Biological Bases of Family Law: Lessons to Be Learned for the Evolutionary Analysis of Law. *International Journal of Law in Context* 2 (3): 277–292.

Cavalli-Sforza, Luigi Luca. 1971. *The Genetics of Human Populations*. San Francisco: Freeman and Co.

Chomsky, Noam. 2002. In *On Nature and Language*, ed. Adriana Belletti and Luigi Rizzi. Cambridge, UK: Cambridge University Press.

Chomsky, Noam, and Michel Foucault. 2006. *The Chomsky-Foucault Debate: On Human Nature*. New York: The New Press.

Ciacci, Margherita. 1983. Significato e interazione: dal behaviorismo sociale all'interazionismo simbolico. In *Interazionismo simbolico*. Bologna: Il Mulino.

Cimatti, Felice. 2013. *Filosofia dell'animalità*. Roma, Bari: Laterza.

Collins, Randall. 1975. *Conflict Sociology. Toward an Explanatory Science*. New York, San Francisco, London: Academic Press.

———. 1988. *Theoretical Sociology*. San Diego: Harcourt Brace Jovanovich.

Conte, Rosaria. 1997. *L'obbedienza intelligente*. Roma, Bari: Laterza.

Crespi, Franco. 1989. *Azione sociale e potere*. Bologna: Il Mulino.

Crippen, Timothy. 1994. Neo-Darwinian Approaches in the Social Sciences: Unwarranted Concerns and Misconceptions. *Sociological Perspectives* 37 (3): 391–401.

Daly, Martin, and Margo Wilson. 1997. Evolutionary Social Psychology and Family Homicide. In *The Maladapted Mind. Classical Readings in Evolutionary Psychopathology*, ed. Simon Baron-Cohen. Hove, New York: Psychology Press.

Damasio, Antonio R. 1994. *Descartes' Error. Emotion, Reason, and the Human Brain*. New York: Putnam's and Sons.

Darwin, Charles. 1859. The Origin of Species.

———. 1872. *The Expression of the Emotions in Man and Animals*. London: Murray.

Dauer, Edward A. 2006. Strong Reciprocity and Accountability: Behavioural Analysis of Patients' Legal Responses to Medical Injury. *International Journal of Law in Context* 2 (3): 257–276.

Dawkins, Richard. 1976. *The Selfish Gene*. New York, Oxford: Oxford University Press.

De Tullio, Elena. 2001. Edward Westermarck e la sua opera. Doctoral Thesis, Università degli Studi di Milano.

de Waal, Frans B.M. 1992. The Chimpanzee's Sense of Social Regularity and Its Relation to the Human Sense of Justice. In *The Sense of Justice. Biological Foundations of Law*, ed. Roger D. Masters and Margaret Gruter. Newbury Park, London, New Dehli: Sage.

———. 2000. The First Kiss. Foundations of Conflict Resolution Research in Animals. In *Natural Conflict Resolution*, ed. Filippo Aureli and Frans B.M. de Waal. Berkeley, Los Angeles: University of California Press.

Dennett, Daniel C. 1996. *Darwin's Dangerous Idea. Evolution and the Meanings of Life*. London: Penguin Books.

Di Giovine, Ombretta. 2009. *Un diritto penale empatico? Diritto penale, bioetica e neuroetica*. Torino: Giappichelli.

Dunbar, R.I.M. 2009. Evolution in Anthropology: A Comment on Nettle's 'Beyond Nature versus Culture'. *Journal of the Royal Anthropological Institute* 15 (2): 245.

Durkheim, Émile. 1893. In *The Division of Labor in Society. [2013]*, ed. Steven Lukes. Basingstoke: Palgrave Macmillan.

Dyevre, Arthur. 2014. Law and the Evolutionary Turn: The Relevance of Evolutionary Psychology for Legal Positivism. *Ratio Juris* 27 (3): 364–386.

Elliott, E. Donald. 1985. The Evolutionary Tradition in Jurisprudence. *Columbia Law Review* 85: 38.

———. 1999. Evolutionary Models in Law: Pros and Cons. In *Law & Evolutionary Biology*, ed. Lawrence A. Frolik, Wolfgang Fikentscher, and Gerti Dieker. Portola Valley: Gruter Institute for Law and Behavioral Research.

Ellis, Lee. 1996. A Discipline in Peril: Sociology's Future Hinges on Curing Its Biophobia. *The American Sociologist* 27 (2): 21–41.
Ferrari, Vincenzo. 1997. *Lineamenti di sociologia del diritto. I. Azione giuridica e sistema normativo*. Roma, Bari: Laterza.
Foster, Kevin R., Katie Parkinson, and Christopher R.L. Thompson. 2007. What Can Microbial Genetics Teach Sociobiology? *Trends in Genetics: TIG* 23 (2): 74–80.
Foucault, Michel. 2004. *Naissance de La Biopolitique: Cours Au College de France (1978–1979)*. Paris: Gallimard.
Frank, Jerome. 1949. *Law and the Modern Mind*. London: Stevens & Sons Limited.
Frank, Robert. 1992. Emotion and the Costs of Altruism. The Economic Foundations of the Sense of Justice. In *The Sense of Justice. Biological Foundations of Law*, ed. Roger D. Masters and Margaret Gruter. Newbury Park, London, New Dehli: Sage.
Freese, Jeremy. 2008. Genetics and the Social Science Explanation of Individual Outcomes. *American Journal of Sociology* 114 (S1): S1–S35.
Freese, Jeremy, Jui-Chung Allen Li, and Lisa D. Wade. 2003. The Potential Relevances of Biology to Social Inquiry. *Annual Review of Sociology* 29: 233–256.
Friedman, Daniel, and Barry Sinervo. 2016. *Evolutionary Games in Natural, Social, and Virtual Worlds*. New York: Oxford University Press.
Gallino, Luciano. 1982. Proprietà dei sistemi nella riproduzione biologica e culturale. In *Sociobiologia possibile. Neodarwinismo e scienze dell'uomo: la ricerca di un'alternativa al determinismo biologico*, ed. Marco Ingrosso, Sergio Manghi, and Vittorio Parisi. Milano: FrancoAngeli.
———. 1987. *L'attore sociale. biologia, cultura e intelligenza artificiale*. Torino: Giulio Einaudi.
Garapon, Antoine. 2012. *Lo stato minimo. Il neoliberalismo e la giustizia*. Milano: Raffaello Cortina.
Gennaro, Giovanni. 2010. Presentazione. In *Neodarwinismo e scienze sociali*, ed. Roberto Vignera. Milano: FrancoAngeli.
Ghezzi, Morris Lorenzo. 1987. *Devianza tra fatto e valore nella sociologia del diritto*. Milano: Giuffrè.
Gianola, Alberto. 1995. L'analisi etologica del diritto. *Rivista di diritto civile*, no. 6.
———. 1997. Evoluzione e diritto. *Rivista di diritto civile*, no. 4.
Gilson, Ronald J., and Robert H. Mnookin. 1995. Cooperation and Competition in Litigation: Can Lawyers Dampen Conflict? In *Barriers to Conflict Resolution*, ed. Kenneth Arrow, Robert H. Mnookin, Lee Ross, Amos Tversky, and Robert Wilson. New York, London: W.W. Norton & Company.
Gladue, B. 1989. Evolutionary Controversy and Psychobiology. *Journal of Social and Biological Systems* 12 (2–3): 251–258.
Goffman, Erving. 1969. *La vita quotidiana come rappresentazione. Anchor Books*. Bologna: Il Mulino.
Goldberg, Aaron D., C. David Allis, and Emily Bernstein. 2007. Epigenetics: A Landscape Takes Shape. *Cell* 128 (4): 635–638.
Goldman, David. 2012. *Our Genes, Our Choices. How Genotype and Gene Interactions Affect Behavior*. Amsterdam: Academic Press.
Gommer, Hendrik. 2012. The Biological Essence of Law. *Ratio Juris* 25 (1): 59–84.
Greenberg, Jerald, and Ronald L. Cohen. 1982. Why Justice? Normative and Instrumental Interpretations. In *Equity and Justice in Social Behavior*, ed. Jerald Greenberg and Ronald L. Cohen. New Haven, London: Academic Press.
Gruter, Margaret. 1979. The Origins of Legal Behavior. *Journal of Social and Biological Structures* 2: 43–51.
———. 1992. An Ethological Perspective on Law and Biology. In *The Sense of Justice. Biological Foundations of Law*, ed. Roger D. Masters and Margaret Gruter. Newbury Park, London, New Dehli: Sage.
Hamilton, William Donald. 1964a. The Genetical Evolution of Social Behavior. I. *Journal of Theoretical Biology* 7: 1–16.

———. 1964b. The Genetical Evolution of Social Behavior. II. *Journal of Theoretical Biology* 7: 17–52.

Heil, Reinhard, Stefanie B. Seitz, Harald König, and Jürgen Robienski. 2017. *Epigenetics: Ethical, Legal and Social Aspects*. Wiesbaden: Springer.

Hirshleifer, Jack. 1980. Privacy: Its Origin, Function and Future. *Journal of Legal Studies* 9 (4): 649–664.

Hodgson, Geoffrey M., and Thorbjørn Knudsen. 2010. *Darwin's Conjecture. The Search for General Principles of Social & Economic Evolution*. Chicago, London: University of Chicago Press.

Hoffman, Morris B. 2006. The Neuroeconomic Path of the Law. In *Law and the Brain*, ed. Semir Zeki and Oliver R. Goodenough. Oxford: Oxford University Press.

Homans, George C. 1961. *Social Behavior: Its Elementary Forms*. New York, Burlingame: Harcourt, Brace & World.

Hopcroft, Rosemary L. 2009. The Evolved Actor in Sociology. *Sociological Theory* 27 (4): 390–406.

Hume, David. 1739. In *A Treatise of Human Nature. [1896]*, ed. L.A. Selby-Bigge. Oxford: Clarendon Press https://www.files.ethz.ch/isn/125487/5010_Hume_Treatise_Human_Nature.pdf.

Ingrosso, Marco, and Sergio Manghi. 1982. Serve la sociobiologia? Riflessioni sul dibattito culturale e di opinione in Italia. In *Sociobiologia possibile. Neodarwinismo e scienze dell'uomo: la ricerca di un'alternativa al determinismo biologico*, ed. Marco Ingrosso, Sergio Manghi, and Vittorio Parisi. Milano: FrancoAngeli.

Jablonka, Eva, and Marion J. Lamb. 2007. *L'evoluzione in quattro dimensioni. Variazione genetica, epigenetica, comportamentale e simbolica nella storia della vita*. Torino: UTET.

Jhering, Rudolf von. 1913. In *Law as a Means to An End*, ed. Isaac Husik. Boston: The Boston Book Company.

Jones, Owen D. 1999. Law, Emotions, and Behavioral Biology. In *Law & Evolutionary Biology*, ed. Lawrence A. Frolik, Wolfgang Fikentscher, and Gerti Dieker. Portola Valley: Gruter Institute for Law and Behavioral Research.

———. 2006. Law, Evolution and the Brain: Applications and Open Questions. In *Law and the Brain*, ed. Semir Zeki and Oliver R. Goodenough. Oxford: Oxford University Press.

Jones, Owen D., Robert Ellickson, Ira Ellman, Adam Gifford, Oliver Goodenough, Lydia Jones, Dennis Karjala, et al. 2001. Time-Shifted Rationality and the Law of Law's Leverage: Behavioral Economics Meets Behavioral Biology. *Northwestern University Law Review* 95 (4): 1141–1206.

Jones, Owen D., and Timothy H. Goldsmith. 2005. Law and Behavioral Biology. *Columbia Law Review* 105: 405–502.

Jörn, Walter, and Anja Hümpel. 2017. Introduction to Epigenetics. In *Epigenetics: Ethical, Legal and Social Aspects*, ed. Reinhard Heil, Stefanie B. Seitz, Harald König, and Jürgen Robienski. Wiesbaden: Springer.

Kaati, G., L.O. Bygren, and S. Edvinsson. 2002. Cardiovascular and Diabetes Mortality Determined by Nutrition During Parents' and Grandparents' Slow Growth Period. *European Journal of Human Genetics* 10 (11): 682–688.

Keller, A.G. 1919. Law in Evolution. *Yale Law Journal* 28: 769–783.

Kuhn, Steven. 2017. Prisoner's Dilemma. *The Stanford Encyclopedia of Philosophy*. https://plato.stanford.edu/archives/spr2017/entries/prisoner-dilemma/.

Kull, Kalevi. 2016. What Kind of Evolutionary Biology Suits Cultural Research? *Sign Systems Studies* 44 (4): 634–647.

Laland, Kevin N., and Gillian R. Brown. 2011. *Sense and Nonsense. Evolutionary Perspectives on Human Behavior*. Oxford: Oxford University Press.

Landecker, Hannah, and Aaron Panofsky. 2013. From Social Structure to Gene Regulation, and Back: A Critical Introduction to Environmental Epigenetics for Sociology. *Annual Review of Sociology* 39 (1): 333–357.

Leiter, Brian, and Michael Weisberg. 2009. Why Evolutionary Biology Is (so Far) Irrelevant to Legal Regulation. *Law and Philosophy* 29 (1): 31–74.

Leonard, Anne S., and Jacob S. Francis. 2017. Plant–Animal Communication: Past, Present and Future. *Evolutionary Ecology* 31 (2): 143–151.

Lopreato, Joseph, and F. Penny Anthon Green. 1990. The Evolutionary Foundations of Revolution. In *Sociobiology and Conflict. Evolutionary Perspectives on Competition, Cooperation, Violence and Warfare*, ed. Johan M.G. van der Dennen and Vincent S.E. Falger. London, New York, Tokyo, Melbourne, Madras: Chapman and Hall.

Lucchini, Mario, Sara Della Bella, and Maurizio Pisati. 2013. The Weight of the Genetic and Environmental Dimensions in the Inter-Generational Transmission of Educational Success. *European Sociological Review* 29 (2): 289–301.

Lumsden, Charles J., and Edward O. Wilson. 1981. *Genes, Mind, and Culture. The Coevolutionary Process*. Cambridge, MA, London: Harvard University Press.

Lytton, Hugh, and David M. Romney. 1991. Parents' Differential Socialization of Boys and Girls: A Meta-Analysis. *Psychological Bulletin* 109 (2): 267–296.

Machalek, Richard, and Michael W. Martin. 2004. Sociology and the Second Darwinian Revolution: A Metatheoretical Analysis. *Sociological Theory* 22 (3): 455–476. https://doi.org/10.2307/3648927.

Mäki, Uskali, Adrian Walsh, and Manuela Fernández Pinto, eds. 2018. *Scientific Imperialism. Exploring the Boundaries of Interdisciplinarity*. London, New York: Routledge.

Malinowski, Bronislaw. 1939. The Group and the Individual in Functional Analysis. *American Journal of Sociology* 44 (6): 938–964.

Marra, Realino. 1984. *Il diritto in Durkheim. Sensibilità e riflessione nella produzione normativa*. Saarbrücken: Universität des Saarlandes.

Marraffa, Massimo. 2011. Evoluzione, cognizione e cultura. In *Neuroetica. Scienze del cervello, filosofia e libero arbitrio*, ed. Andrea Lavazza and Giuseppe Sartori. Bologna: Il Mulino.

Maryanski, Alexandra. 1994. The Pursuit of Human Nature in Sociobiology and Evolutionary Sociology. *Sociological Perspectives* 37 (3): 375–389.

Maryanski, Alexandra, and Jonathan H. Turner. 1992. *The Social Cage. Human Nature and the Evolution of Society*. Stanford, CA: Stanford University Press.

Masters, Roger D. 1992a. Naturalistic Approaches to Justice in Political Philosophy and the Life Sciences. In *The Sense of Justice. Biological Foundations of Law*, ed. Roger D. Masters and Margaret Gruter. Newbury Park, London, New Dehli: Sage.

———. 1992b. The Problem of Justice in Contemporary Legal Thought. In *The Sense of Justice. Biological Foundations of Law*, ed. Roger D. Masters and Margaret Gruter. Newbury Park, London, New Dehli: Sage.

Maturana, Humberto R., and Francisco J. Varela. 1980. *Autopoiesis and Cognition. The Realization of the Living*. Dordrecht, Boston: D. Reidel Publishing.

McGuire, Michael T. 1992. Moralistic Aggression, Processing Mechanisms, and the Brain. In *The Sense of Justice. Biological Foundations of Law*, ed. Roger D. Masters and Margaret Gruter. Newbury Park, London, New Dehli: Sage.

Mead, George Herbert. 1972. In *Mind, Self & Society: From the Standpoint of a Social Behaviorist*, ed. Charles W. Morris. Chicago: University of Chicago Press.

Menkel-Meadow, Carrie. 1992. Is Altruism Possible in Lawyering? *Georgia State University Law Review* 8: 385–419.

Mesoudi, Alex, Andrew Whiten, and Kevin N. Laland. 2006. Towards a Unified Science of Cultural Evolution. *Behavioral and Brain Sciences* 29 (4): 329–347.

Miller, Greg. 2010. The Seductive Allure of Behavioral Epigenetics. *Science* 329 (5987): 24–27.

Mormino, Gianfranco. 2016. *Per una teoria dell'imitazione*. Milano: Raffaello Cortina.

Murdock, George Peter. 1945. The Common Denominators of Culture. In *The Science of Man in the World Crisis*, ed. Ralph Linton. New York: Columbia University Press.

Nettle, Daniel. 2009. Beyond Nature Versus Culture: Cultural Variation as an Evolved Characteristic. *Journal of the Royal Anthropological Institute* 15 (2): 223–240.

Nevin, Rick. 2000. How Lead Exposure Relates to Temporal Changes in IQ, Violent Crime, and Unwed Pregnancy. *Environmental Research* 83: 1–22.

———. 2007. Understanding International Crime Trends: The Legacy of Preschool Lead Exposure. *Environmental Research* 104: 315–336.

Nielsen, François. 1994. Sociobiology and Sociology. *Annual Review of Sociology* 20 (January): 267–303.

Nowak, Martin A., Karen M. Page, and K. Sigmund. 2000. Fairness Versus Reason in the Ultimatum Game. *Science* 289 (September): 1773–1775.

Parisi, Vittorio. 1982. Biologia sociale e sociobiologia wilsoniana. In *Sociobiologia possibile. Neodarwinismo e scienze dell'uomo: la ricerca di un'alternativa al determinismo biologico*, ed. Marco Ingrosso, Sergio Manghi, and Vittorio Parisi. Milano: FrancoAngeli.

Pattaro, Enrico. 2005. The Law and the Right. A Reappraisal of the Reality That Ought to Be. In *A Treatise*, ed. Enrico Pattaro. Dordrecht: Springer.

Pennisi, Antonino. 2014. *L'errore Di Platone*. Bologna: Il Mulino.

Pennisi, Elizabeth. 2008. Research Funding. Are Epigeneticists Ready for Big Science? *Science (New York, N.Y.)* 319 (5867): 1177.

Pescosolido, Bernice A., Brea L. Perry, J. Scott Long, Jack K. Martin, John I. Nurnberger Jr., and Victor Hesselbrock. 2008. Under the Influence of Genetics: How Transdisciplinarity Leads Us to Rethink Social Pathways to Illness. *American Journal of Sociology* 114 (S1): S171–S201.

Pinker, Steven. 2002. *The Blank Slate. The Modern Denial of Human Nature*. New York: Penguin Books.

Pitasi, Andrea. 2012. Hypercitizenship and the Management of Genetic Diversity: Sociology of Law and the Key Systemic Bifurcation Between the Ring Singularity and the Neofeudal Age. *World Futures: The Journal of Global Education* 68 (4–5): 314–331.

Pitasi, Andrea, and Mariarosalba Angrisani. 2013. 'Hyper Citizenship' and the Evolution of a Global Identity. *Journal of Sociological Research* 4 (2): 318–335. https://doi.org/10.5296/jsr.v4i2.4369.

Pitasi, Andrea, and Sara Petroccia. 2018. Expanding Citizenship. In *Development as Citizenship Expansion. A Systemic Approach to the Globalizing Law System*, ed. Natália Brasil Dib, André Parmo Folloni, and Sara Petroccia. Curitiba: Editora Íthala.

Pocar, Valerio. 2005. *Gli animali non umani: per una sociologia dei diritti*. Roma: Laterza.

Popper, Karl Raimund. 1971. *The Open Society and Its Enemies. 2. The High Tide of Prophecy: Hegel, Marx and the Aftermath*. Princeton: Princeton University Press.

———. 1978. Natural Selection and the Emergence of Mind. *Dialectica* 32 (3–4): 339–355.

Proctor, Darby, Rebecca A. Williamson, Frans B.M. de Waal, and Sarah F. Brosnan. 2013. Chimpanzees Play the Ultimatum Game. *Proceedings of the National Academy of Sciences of the United States of America* 110 (6): 2070–2075.

Rabinow, Paul. 1996. Artificiality and Enlightenment: From Sociobiology to Biosociality. In *Essays on the Anthropology of Reason*, 91–111. Princeton: Princeton University Press.

Raine, Adrian. 2014. *The Anatomy of Violence. The Biological Roots of Crime*. New York: Vintage Books.

Rand, David G., and Martin A. Nowak. 2011. The Evolution of Antisocial Punishment in Optional Public Goods Games. *Nature Communications* 2 (January): 434.

Ravelli, G.P., Z.A. Stein, and M.W. Susser. 1976. Obesity in Young Men After Famine Exposure in Utero and Early Infancy. *The New England Journal of Medicine* 295 (7): 349–353.

Resta, Eligio. 2002. *Il diritto fraterno*. Roma, Bari: Laterza.

Richerson, Peter J., and Robert Boyd. 1999. Complex Societies. The Evolutionary Origins of a Crude Superorganism. *Human Nature* 10 (3): 253–289.

———. 2001. Built for Speed, Not for Comfort. Darwinian Theory and Human Culture. *History and Philosophy of the Life Sciences* 23: 425–465.

Ridley, Matt. 2003. *Nature Via Nurture: Genes, Experience, and What Makes Us Human*. New York: HarperCollins.

Robienski, Jürgen. 2017. Epigenetics and Legal Regulations: A Challenge for the State's Obligation to Protect Environment, Individual's Health and Civil Rights. In *Epigenetics: Ethical, Legal and Social Aspects*, ed. Reinhard Heil, Stefanie B. Seitz, Harald König, and Jürgen Robienski. Wiesbaden: Springer.

Robinson, Gene E., Christina M. Grozinger, and Charles W. Whitfield. 2005. Sociogenomics: Social Life in Molecular Terms. *Nature Reviews. Genetics* 6 (4): 257.

Rose, Hillary, and Steven Rose. 2014. *Geni, cellule e cervelli. Speranze e delusioni della nuova biologia*. Torino: Codice Edizioni.

Rose, Nikolas. 2007. *The Politics of Life Itself: Biomedicine, Power, and Subjectivity in the Twenty-First Century*. Princeton, Oxford: Princeton University Press.

Roseboom, T.J., J.H. van der Meulen, C. Osmond, D.J. Barker, A.C. Ravelli, J.M. Schroeder-Tanka, G.A. van Montfrans, R.P. Michels, and O.P. Bleker. 2000. Coronary Heart Disease after Prenatal Exposure to the Dutch Famine, 1944–45. *Heart (British Cardiac Society)* 84 (6): 595–598.

Rossetti, Carlo Giuseppe. 1982. Può la biologia essere una scienza sociale? In *Sociobiologia possibile. Neodarwinismo e scienze dell'uomo: la ricerca di un'alternativa al determinismo biologico*, ed. Marco Ingrosso, Sergio Manghi, and Vittorio Parisi. Milano: FrancoAngeli.

Rossi, Alice S. 1984. Gender and Parenthood. *American Sociological Review* 49 (1): 1–19.

Rothstein, Mark A., Yu Cai, and Gary E. Marchant. 2009. The Ghost in Our Genes: Legal and Ethical Implications of Epigenetics. *Health Matrix* 19 (February): 1–62.

Rottleuthner, H. 1985. Biologie Und Recht. *Zeitschrift Fur Rechtsoziologie* 6: 104–126.

Rozin, Paul, Jonathan Haidt, and Clark R. McCauley. 2009. Disgust. In *Oxford Companion to Affective Sciences*, ed. David Sanders and Klaus Scherer. Oxford: Oxford University Press.

Runciman, Walter G. 1970. *Sociology in Its Place*. Cambridge, UK: Cambridge University Press.

———. 1999. *The Social Animal*. London: Fontana.

———. 2005. Stone Age Sociology. *The Journal of the Royal Anthropological Institute* 11 (1): 129–142.

———. 2009. *The Theory of Cultural and Social Selection*. Cambridge, UK: Cambridge University Press.

Rusconi, Francesco, Barbara Grillo, Luisa Ponzoni, Silvia Bassani, Emanuela Toffolo, Leda Paganini, Alessandra Mallei, et al. 2016. LSD1 Modulates Stress-Evoked Transcription of Immediate Early Genes and Emotional Behavior. *Proceedings of the National Academy of Sciences of the United States of America* 113 (13): 3651–3656.

Ruse, Michael. 1979. *Sociobiology: Sense or Nonsense*. Dordrecht, Boston, London: R. Reidel Publishing Company.

Sacco, Rodolfo. 2015. *Il diritto muto*. Bologna: Il Mulino.

Salvadori, Guglielmo. 1904. L'idea del diritto e della giustizia nella filosofia dell'evoluzione. *Rivista italiana di sociologia* 8 (1).

Sampson, Edward E. 1983. *Justice and the Critique of Pure Psychology*. New York, London: Plenum Press.

Sandi, Carmen, and József Haller. 2015. Stress and the Social Brain: Behavioural Effects and Neurobiological Mechanisms. *Nature Reviews Neuroscience* 16 (5): 290–304.

Sarti, Simone. 2015. *Evoluzione e complessità sociale. Introduzione a una teoria del cambiamento sociale*. Novara: De Agostini.

Scherer, Klaus R. 1992. Issues in the Study of Justice. In *Justice. Interdisciplinary Perspectives*, ed. Klaus R. Scherer. Cambridge, UK: Cambridge University Press.

Schnittker, Jason. 2008. Happiness and Success: Genes, Families, and the Psychological Effects of Socioeconomic Position and Social Support. *American Journal of Sociology* 114 (S1): S233–S259.

Schuol, Sebastian. 2017. Epigenetics and Genetic Determinism (in Popular Science). In *Epigenetics: Ethical, Legal and Social Aspects*, ed. Reinhard Heil, Stefanie B. Seitz, Harald König, and Jürgen Robienski. Wiesbaden: Springer.

Searle, John R. 2001. *Rationality in Action*. Cambridge, MA: The MIT Press.

Segerstråle, Ullica. 1990. The Sociobiology of Conflict and the Conflict About Sociobiology: Science and Morals in the Larger Debate. In *Sociobiology and Conflict. Evolutionary Perspectives on Competition, Cooperation, Violence and Warfare*, ed. Johan M.G. van der Dennen and Vincent S.E. Falger. London, New York, Tokyo, Melbourne, Madras: Chapman and Hall.

———. 2000. *Defenders of the Truth. The Battle for Science in the Sociobiology Debate and Beyond*. Oxford, New York: Oxford University Press.

Shanahan, Michael J., Stephen Vaisey, Lance D. Erickson, and Andrew Smolen. 2008. Environmental Contingencies and Genetic Propensities: Social Capital, Educational Continuation, and Dopamine Receptor Gene DRD2. *American Journal of Sociology* 114 (S1): S260–S286.

Sidoti, Francesco, and Mariateresa Gammone. 2010. The Lombroso Biologism: A Centenary of Controversies. *Salute e società* 3: 114–128.

Simmel, Georg. 1927. *Frammento Sull'amore*. Milano: Edizioni Athena.

Singer, Peter. 2000. *A Darwinian Left. Politics. Evolution and Cooperation*. New Haven and London: Yale University Press.

Slesinger, Donald, and Robert M. Hutchins. 1929. Some Observations on the Law of Evidence: Family Relations. Paper 4839. *Yale Law School. Faculty Scholarship Series*.

Smith, Thomas S., and Gregory T. Stevens. 1998. Biological Foundations of Social Interaction. Computational Explorations of Nonlinear Dynamics in Arousal-Modulation. In *Chaos, Complexity, and Sociology. Myths, Models, and Theories*, ed. Raymond A. Eve, Sarah Horsfall, and Mary E. Lee. Thousand Oaks, London, New Dehli: Sage.

Spencer, Herbert. 1898. *The Principles of Sociology. [2002]*. New Brunswick (NJ), London: Transaction Publishers.

Stake, Jeffrey Evans. 2006. The Property 'Instinct'. In *Law and the Brain*, ed. Semir Zeki and Oliver R. Goodenough. Oxford: Oxford University Press.

Stewart-Williams, Steve. 2010. *Darwin, God, and the Meaning of Life: How Evolutionary Theory Undermines Everything You Thought You Knew*. Cambridge, UK: Cambridge University Press.

Symons, Donald. 1992. On the Use and Misuse of Darwinism in the Study of Human Behavior. In *The Adapted Mind. Evolutionary Psychology and the Generation of Culture*, ed. Jerome H. Barkow, Leda Cosmides, and John Tooby. New York, Oxford: Oxford University Press.

Tarde, Gabriel. 1903. In *The Laws of Imitation*, ed. Elsie Clews Parsons. New York: Henry Holt & Company.

Thaler, Richard H., and Cass R. Sunstein. 2009. *Nudge. Improving Decisions About Health, Wealth and Happyness*. London: Penguin Books.

Tooby, John, and Lela Cosmides. 1992. The Psychological Foundations of Culture. In *The Adapted Mind. Evolutionary Psychology and the Generation of Culture*, ed. Jerome H. Barkow, Leda Cosmides, and John Tooby. New York, Oxford: Oxford University Press.

Trivers, Robert. 2002. *Natural Selection and Social Theory: Selected Papers of Robert Trivers*. Oxford: Oxford University Press.

Tung, Jenny, Luis B. Barreiro, Zachary P. Johnson, Kasper D. Hansen, Vasiliki Michopoulos, Donna Toufexis, Katelyn Michelini, Mark E. Wilson, and Yoav Gilad. 2012. Social Environment Is Associated with Gene Regulatory Variation in the Rhesus Macaque Immune System. *Proceedings of the National Academy of Sciences of the United States of America* 109 (17): 6490–6495.

Turillo, Carmelo Joseph, Robert Folger, James J. Lavelle, Elizabeth E. Umphress, and Julie O. Gee. 2002. Is Virtue Its Own Reward? Self-Sacrificial Decisions for the Sake of Fairness. *Organizational Behavior and Human Decision Processes* 89 (1): 839–865.

van den Berghe, Pierre L. 1990. Why Most Sociologists Don't (And Won't) Think Evolutionarily. *Sociological Forum* 5 (2): 173–185.

van Schaik, Carel P., and Michel Kai. 2016. *The Good Book of Human Nature. An Evolutionary Reading of the Bible*. New York: Basic Books.

Veenendaal, M.V.E., R.C. Painter, S.R. de Rooij, P.M.M. Bossuyt, J.A.M. van der Post, P.D. Gluckman, M.A. Hanson, and T.J. Roseboom. 2013. Transgenerational Effects of Prenatal Exposure to the 1944–45 Dutch Famine. *BJOG: An International Journal of Obstetrics & Gynaecology* 120 (5): 548–554.

Vezzoni, Paolo. 2010. Determinismo genetico: considerazioni originate dalle patologie ereditarie. In *Neodarwinismo e scienze sociali*, ed. Roberto Vignera. Milano: FrancoAngeli.

Vignera, Roberto. 2010. Modelli evolutivi e teoria sociologica. In *Neodarwinismo e scienze sociali*, ed. Roberto Vignera. Milano: FrancoAngeli.

Vineis, Paolo. 2006. *Equivoci bioetici*. Torino: Codice Edizioni.

von Hayek, Friedrich August. 2013. *Law, Legislation and Liberty: A New Statement of the Liberal Principles of Justice and Political Economy*. London, New York: Routledge.

Waddington, Conrad Hal. 1942. Canalization of Development and the Inheritance of Acquired Characters. *Nature* 150 (3811): 563.

Wastell, David, and Sue White. 2017. *Blinded by Science. The Social Implications of Epigenetics and Neuroscience*. Bristol: Policy Press.

Weaver, Ian C.G., Nadia Cervoni, Frances A. Champagne, Ana C. D'Alessio, Shakti Sharma, Jonathan R. Seckl, Sergiy Dymov, Moshe Szyf, and Michael J. Meaney. 2004. Epigenetic Programming by Maternal Behavior. *Nature Neuroscience* 7 (8): 847.

Wentworth, William, and Darrell Yardley. 1994a. Deep Sociality: A Bioevolutionary Perspective on the Sociology of Emotions. In *Social Perspectives on Emotion*, ed. William M. Wentworth and John Ryan. Bingley: Emerald.

———. 1994b. Introduction. In *Social Perspectives on Emotion*, ed. William Wentworth and Darrell Yardley. Bingley: Emerald.

Wilkinson, Richard, and Michael Marmot. 2003. *Social Determinants of Health Inequalities. The Solid Facts*. World Health Organization: http://www.euro.who.int/__data/assets/pdf_file/0005/98438/e81384.pdf.

Wilson, David Sloan. 2015. *Does Altruism Exist? Culture, Genes, and the Welfare of Others*. New Haven: Yale University Press.

Wilson, Edward O. 2000. *Sociobiology. The New Synthesis*. Cambridge, MA, London: The Bellknap Press of Harvard University Press.

Wrangham, Richard. 2006. Why Apes and Humans Kill. In *Conflict*, ed. Martin Jones and A.C. Fabian. Cambridge: Cambridge University Press.

Wright, Robert. 1995. *The Moral Animal: Why We Are The Way We Are*. New York: Vintage.

Yehuda, Rachel, Nikolaos P. Daskalakis, Linda M. Bierer, Heather N. Bader, Torsten Klengel, Florian Holsboer, and Elisabeth B. Binder. 2016. Holocaust Exposure Induced Intergenerational Effects on FKBP5 Methylation. *Biological Psychiatry* 80 (5): 372–380.

Yoneya, Kinuyo, and Junji Takabayashi. 2014. Plant–Plant Communication Mediated by Airborne Signals: Ecological and Plant Physiological Perspectives. *Plant Biotechnology* 31 (5): 409–416.

Zak, Paul J. 2006. Neuroeconomics. In *Law and the Brain*, ed. Semir Zeki and Oliver R. Goodenough. Oxford: Oxford University Press.

Zamboni, Mauro. 2010. Evolutionary Theory and Legal Positivism: A Possible Marriage. Vol. 306. http://ssrn.com/abstract=1657841.

Zippelius, Reinhold. 1986. Exclusion and Shunning as Legal and Social Sanctions. *Ethology and Sociobiology* 7 (3–4): 159–166.

Chapter 4
Cognition and the Law

4.1 The Brain, Cognitive Science, and Law

With the demise of nineteenth-century behavioral mechanicism came the need to chart a new course in the effort to work toward a modern theory of behavior, and so also of law-oriented behavior, and it was the whole set of mentalist and behaviorist conceptions that pointed the way forward. Ever since the Humean intuition that the passions dominate over rationality and propel the will, the idea emerged that pure subjectivity could explain much more than just its own capture of reason, but this idea couldn't be fully developed until the analysis of mental functions found at its disposal the full arsenal of the contemporary neurosciences (Oliverio 2012) and cognitive science (Williams 2001). Neuroscience and the cognitive sciences (Anolli and Mantovani 2012, p. 84) changed the paradigm in the social sciences, too, because they showed how it was possible to bridge the gap between biology and culture, matter and mind (Pinker 2002, pp. 34, 51).

Neuroscience is an inherently interdisciplinary science focused on the analysis of the nervous system, and placed at the intersection of medicine, biology, physics, and psychology. The cognitive sciences, rather than a discipline with established boundaries, constitute an interdisciplinary approach to the study of thinking systems (natural or artificial) that draws not only on neuroscience but also on psychology, philosophy, linguistics, and artificial intelligence (Miller 1956; Nagel 1974).

In vivo brain study techniques have identified how emotions and perceptions have a physical substratum consisting of neuronal activity and of the "biological machinery of reason" (Damasio 1994, p. 53). In this, neuroscience has provided the cognitive sciences with an indispensable contribution: every part of the body can reach the brain through the signals sent out by the nervous system and the chemicals produced by the body's activity (Almodóvar 2015). The brain, on the other hand, is capable of acting on the whole body through the nerves, making up an undistinguishable whole (Damasio 1994, p. 88). The neurosciences have also made it possible to locate in the prefrontal cortex the evolutionarily determinant seat of the

L. Cominelli, *Cognition of the Law*,
https://doi.org/10.1007/978-3-319-89348-8_4

development of the higher functions, something that physiologically resembles the *superego* (Sapolsky 2006, p. 233).

However, the deterministic analogies stop here: there is no area of the brain toward which all the sensations converge (no "Cartesian theater"), nor is there a "command center" where decisions are taken (Damasio 1994, p. 94). If we look at the physiological evolution of the brain—from the reflexes of the primitive spinal cord to the instincts and pulsions of the paleoencephalon, culminating in the cognitive activities of the neoencephalon—we will not find much linearity (Oliverio 2012, p. 68). The brain is not a homogeneous, undifferentiated organ but a complex one, made up of regions whose mutual interactions are not yet fully understood (Greenberg and Cohen 1982, p. 81). Just think of the distinct functional areas that were thought to be rather rigidly attributable to the right or left hemisphere: this hypothesis has already been disproven by research on the subjects in whom the two hemispheres were anatomically separated (Gazzaniga 1985, p. 47). Rationality, which is thought to reside on the neocortical level, would seem to be useless without the biological regulation provided by the subcortical areas, and would therefore be activated only through concerted activity with other areas of the brain, such as the amygdala or the cingulate cortex front (Damasio 1994, pp. 128, 137).

Damasio's work highlights the fundamental defect of the Cartesian assumption, consisting in leaving the emotions and feelings outside the cognitive framework, and postulating a dualism between a *res cogitans* and a *res extensa*. Once the biological origin of affective states is correctly understood, this is no longer possible. The linking, as Damasio's hypothesis goes, is carried out by somatic markers which simplify the decision-making activity, and which in essence represent particular emotions generated by secondary emotions. The markers, subject to learning, are linked to predictable developments of certain future scenarios, acting as alarms bells through negative markers or as incentives through positive markers (Damasio 1994, p. 173ff.). Social markers are physiological mechanisms that are dependent on the autonomic nervous system (perspiration, cardiac acceleration, muscle contraction, gastrointestinal contractions) and indirectly guide our rational choices. In explaining unconscious phenomena, the Freudian unconscious, linked to repressed desires and childhood experiences, should be complemented or even replaced by a cognitive unconscious (Oliverio 2012, p. 66).

This cognitive revolution has remarkable implications for all of the social sciences. If cognitive psychology and psychoanalysis have helped to undermine *homo oeconomicus* (Crespi 1989, p. 30), we cannot expect *homo sociologicus* to remain unscathed. The complex social interactions developed by primates and then by human beings—interactions such as cooperation, imitation, or deception—are precisely what would favor the impetuous expansion of the cerebral cortex (Pineda 2009, p. v). In addition to challenging the psychological and sociological paradigms, the cognitive sciences overcome the Cartesian error and the mind-body dichotomy, and in this way they lay the groundwork for a conceptual convergence of the philosophical, anthropological, and pedagogical sciences (Iagulli 2011, p. 13).

In the coming decades, according to Pattaro (2005, p. 397), the cognitive sciences and the neurosciences will contribute to developing those fields of investigation in which psychology and sociology often proceed "in a more or less figurative way." With the newly gained knowledge of individual cognitive mechanisms, the phenomenological tradition (into which I include methodological individualism) could make headway in understanding some major macrosociological questions that have been left unresolved. Collins (1975, p. 14) exemplifies the potentialities of the cognitive approach by reference to Weber's concept of the legitimization of power and to Durkheim's concept of the production of a collective consciousness through rituals. The relevance of the mind's physiological mechanisms in social experience and behavior is clearly also asserted by Mead (1972), just as unrecognized references to neuroscience are clearly highlighted in the symbolic interactionist tradition (Franks 2010, p. 2).

With the convergence of these paradigms, the traditional mainstay disciplines (ethics, sociology, economics) have come to find a common underpinning in "neuro-," giving us neuro-ethics, neuro-sociology, and neuro-economics (Boella 2011; Franks 2010; Hitlin and Vaisey 2013). This is prompting an effort to retest some past assumptions by introducing the neuro-cognitive variable. Thus, we have discovered that we possess an unconscious ethic that leads us to make tragic choices in a reflexive manner, sometimes with paradoxical outcomes. Think of the well-known ethical conundrum posed by the trolley problem, which sets up a hypothetical situation where a runaway trolley is going to run over and kill either one person or several: we are in a position to decide that outcome by pulling a switch that would turn the trolley onto a dead track, thus killing one person (standing on the dead track) in order to save several (on the main track), and we would easily agree that this is the right thing to do. But change the method by which to achieve the same outcome—by instead pushing the one man onto the track so as to stop the trolley that is heading for the several—and we are no longer so sure that we want to do that, even if there is no apparent moral distinction that can be appreciated by comparison with the first scenario (Pardo and Patterson 2013, p. 54).

Given the comprehensible objections to the view that individual experience is entirely unique and irreducible (Manzotti and Tagliasco 2008), it has been emphasized that using empirical data is not incompatible with asking "properly ethical questions about beliefs, choices and their justifications, values, and life projects and their historical and cultural variations" (Boella 2011, p. 89), or with approaching "ethics as an empirico-experimental discipline, contiguous with medicine" (ibid., p. 96).

Into the neuro-social sciences comes *neurolaw* (Alces 2018; Jones et al. 2014; Fuselli 2014; Picozza 2013; Picozza and D'Arrigo 2015). The object of neurolegal analysis is to identify the neurological basis of law-oriented behavior, and to integrate the cognitive sciences into legal knowledge, so as to solve some problems such as the legal capacity to exercise rights, as well as criminal responsibility, antisocial behavior, and the reliability of witnesses (Palazzani and Zanotti 2013, pp. 4–6). Just like the other social sciences, the theory of law and the sociological analysis of law are looking for cognitive models to explain behavior, in such a way as to complement the traditional "instrumental" perspectives of rational or normative-moral choice (Suchman 1997, pp. 476, 482). Cognitive mechanisms are recognized as indispensable to an understanding of social phenomena such as law, with the aim of overlaying the legal validity of rules onto the properties that make them acceptable or worthy and cognitively binding (Sartor 2005, p. 331).

When neurolaw was first developed in the 1980s, its interests were strictly theoretical. Then it focused on the limited fields of criminal imputability and mental defect (Greely 2004, pp. 141–142; Goodenough and Tucker 2010, p. 80), and began to gradually influence court decisions. Since the 1990s, that influence has essentially been spreading throughout the legal system (Santosuosso 2009, p. 13), at first in criminal law and then across a whole range of other areas, such as investor and consumer protection (Monti 2008), and indeed wherever cognitive distortions are likely to lead to substantial injustice.

Individual perceptions and judgments (Goodenough 2006, p. 100) are evidently also a question that legal science has not sufficiently investigated, and where the wisdom of the jurist has proved insufficient. Why, for example, are intellectual property infringements not perceived as such by their perpetrators (Goodenough and Prehn 2006; Goodenough 2009), and why do lawmakers commit legislative failures by passing unenforceable laws (Monti 2008, p. 16)? Neurolaw could also support the assessment of neuro-cognitive enhancement (Fuselli 2014; Garland 2004a; Goodenough and Tucker 2010; Morse 2004; Palazzani 2013), meaning the medical enhancement of mental performance in the absence of disease, and so without a real therapeutic need.

As noted, however, cognitive neuroscience and neurolaw are inherently transdisciplinary and would not survive any disciplinary barriers: they cannot be conceived as independent disciplines (Caterina 2008a, p. 1). What the neuro-legal approach can contribute to the study of law is the ability to overcome the nomothetic orientation of legal theory, which proceeds from the assumption that all individuals have analogous cognitive abilities (a tenet in behavioral law and economics, too). This approach can thus help us recognize that reality is instead invariably idiographic, and that the law should whenever possible take account of the different individual cognitive faculties by way of ad hoc mechanisms (Rachlinski 1982, p. 126). The justification offered by those who argue in favor of nomothetic models for lawmaking purposes can be compared to the justification for rational choice theory, which is that these models would offer an approximation that overall is satisfactory, given that cognitive errors are for the most part poorly recognizable, and small enough to be of negligible importance (ibid., p. 139).

The neuro-approach in the social sciences does not require that human behavior be explained by reducing it to the level of neural procedures, but it does, after all, involve a leap: mental functions, which we intuitively consider to be indistinctly performed, need to be seen modularly. The modular vision of the mind (Fodor 1983), endorsed by evolutionary psychology (Barkow et al. 1992), describes the higher functions as divisible in "expert systems" of simulations, "tasked with representing and 'treating' at a mental level problems that are related to specific sectors of the environment" (Anolli and Mantovani 2012, p. 89).

Mental modules offer a reliable map of the world and are the basis of success in dealing with hostile situations. They are inherited from generation to generation, and therefore acquire a universal, fixed, and biologically determined aspect (ibid., p. 90).

> What is special about the human mind is not that it gave up instinct in order to become flexible, but that it proliferated instincts—that is, content-specific problem-solving specializations—which allowed an expanding role for psychological mechanisms that are (relatively) more function-general. These are presently lumped into categories with unilluminating labels such as "the capacity for culture," "intelligence," "learning," and "rationality." (Tooby and Cosmides 1992, p. 113)

Morals and law, too, may be amenable to distinct brain functions, and it has been specifically hypothesized that justice and law depend on separate mental functions: justice corresponds to a "nonverbal algorithm that is programmed by some mixture of genetic blueprint, cultural heritage and personal experience," while law, on the contrary, is founded on a system of thought based on language, and so can be referred to as an "interpreter module" (Pardo and Patterson 2013, p. 50). But it is difficult to test this hypothesis, since we lack sufficiently sophisticated techniques and face the challenge of unambiguously linking distinct mental functions to the activation of circumscribed brain areas.

Neurolaw is primarily engaged in a debate that will be explored further on (in Sect. 4.2), regarding legal capacity and individual imputability, but we need to address one of the most pressing issues that derive from the deterministic view of the brain in relation to the problem of individual conscience and freedom of choice. Neuroethics has attempted to solve this problem primarily by the distinguishing between freedom of choice and freedom of action, and the issue has already been extensively addressed in relation to the implications it carries for legal responsibility (Cary 2007; Fuselli 2013; Kawohl and Habermeyer 2007; Kroeber 2007; Morse 2004).

The debate on determinism and free will is still too fluid to provide any firm anchor. A widely discussed thesis is that the law assumes an overly ambitious notion of free will and ignores the deterministic criticisms of that notion formulated in cognitive neuroscience (Green and Cohen 2006, p. 208). The neuroscientific investigation of free will changed radically after Libet's experiments on cerebral electrical impulses preceding voluntary acts (Libet et al. 1983). Although it would later be questioned whether these experiments were conclusive, the study did seem to say that consciousness followed action, suggesting that consciousness is in a sense a spectator of acts previously decided, deterministically, by another cerebral function (Sartori et al. 2011, p. 158). In the experiment, subjects were shown a movement of the finger that they were then asked to repeat spontaneously, that is, of their own volition and when they felt the impulse to do so, but indicating the impulse to the experimenters the moment it came; in this way, the experimenters could measure the time when the brain sent the command signal to move the finger relative to the time when the subjects indicated that they felt the impulse to do so. It was determined that the impulse came about 20 hundredths of a second before carrying out the movement. However, 55 hundredths of a second before the subjects became aware of the impulse (that is, before signaling it to the experimenters), the electrical activity in their brains surged, a phenomenon termed "readiness potential." This was interpreted as evidence that the movement may not be imputable to the will, but to a previous neural event, outside any possible conscious control. As noted, however, this interpretation was challenged, since the movement had been primed by an external instruction. Other experiments were subsequently carried out to further investigate this research hypothesis in a more sophisticated way (Pockett 2007, p. 290), but they could not rebut the objection that mental events cannot be considered free decisions in a subjective sense, and perhaps are not even true decisions (De Caro 2011, pp. 75–76).

In the study of consciousness and of our awareness of our own actions, with their obvious repercussions at the level of legal attitudes, different conceptions have coalesced around three fundamental views in the philosophy of the mind (Farahany 2012). The first of these is the *materialistic-deterministic* view, which accepts the substance of Libet's experiments, and of subsequent ones on the "readiness potential," and maintains that mental events cannot escape the physical laws that regulate the processes of matter, which is equivalent to rejecting free will. In the behavioral sciences, this view is characterized as "incompatibilist" (Pockett 2007, p. 282). Pockett tries to differentiate this position from pure determinism by arguing that conscience coincides not with the brain but with a "field" generated by the brain. This makes it possible to preserve the empirical verifiability of the hypothesis.

Next we have the *dualist* or *libertarian* view, on which neurological phenomena and conscious volition necessarily operate on two different levels. Volition would be the manifestation of a "neurophysiological indeterminism" that probably finds its justification in quantum indeterminism. This view posits the incommensurability of moral conscience with neural consciousness and has also been labeled "mysterianism." Free will is an indispensable element but is at the same time inexplicable and incompatible with the scientific vision of the world (De Caro 2011, p. 81).

Finally, we have the *compatibilist* view, which claims that neurological determinism can coexist with free will. This hypothesis avoids the reductionist trap, on the reasoning that even if the biological makeup of our brain does not give us full freedom of choice about desires, preferences, and drives, we still retain at least *some* freedom of action (Farahany 2012, p. 2). This freedom specifically consists in the power to inhibit an act that we would otherwise be compelled to carry out under an uncontrollable impulse. So, while free will in its fullest sense would be an illusion, we would still have a "veto power" (Palazzani and Zanotti 2013, p. 2). Even if neural activity precedes awareness, we still retain some flexibility in choosing a course of action, and this shows that there is, after all, a role for conscious intention: the neural activity that precedes awareness would offer a set of alternatives, however limited they may be, making it possible for us to disengage from the chain of events that on the basis of Libet's studies would seem unavoidable (Farahany 2012, pp. 11–12).

The compatibilist stance is embraced by Searle. While he qualifies consciousness as "an ordinary biological phenomenon comparable with digestion or growth" (Searle 1997, p. 6), dependent on a brain that is "a conscious biological machine" (ibid., p. 202), he also underscores its being distinctive in that it cannot be conceived from a "third person" point of view: consciousness and our experience of it form an indistinguishable unity (ibid., p. 213). That consciousness cannot be reduced to neuronal activity, however, should not be taken to mean that we must posit a dualism between the two, with all the consequences that dualism would like to extract from such irreducibility (ibid., p. 214). Although the jury is still out on this debate, the compatibilist view seems to have gained the upper hand today.

The issues still debated in neuroethics and neurolaw have an immediate appeal for sociology as well (Franks 2010), even if some of these discussions seem to take place on an almost metaphysical plane, behind invisible and unmentionable taboos. There is also in this dialogue a roadblock that has been correctly highlighted by Ghezzi (2016), noting that while determinism remains refutable and falsifiable in keeping with the canons of scientificity, this does not obviously apply to indeterminism. The intent of my work is instead to lay out a field of interdisciplinary comparison on testable hypotheses. It is for this reason that we must now address the issue of the use made of neuroscientific findings in the interpretation and enforcement of legal rules.

4.2 Neuroscientific Techniques Applied to the Law

It was previously mentioned that the first legal use of neuroscience was in criminal law: it served as a tool by which to gain a deeper understanding of criminal responsibility and to accordingly redesign criminal proceedings, and this still remains its primary use in the law (Blumoff 2009; Glannon 2014; Macioce 2013; Morse 2009; Pardo and Patterson 2013; Spranger 2009; Basile and Vallar 2017). The responsibility that comes into question in criminal trials has encouraged experimentation with innovative and nonmainstream methods. Thus neuroscience has been used to hone our understanding of the rationality of potential criminals, and this is only beneficial if it can help us to fashion the law so as to make for more effective deterrence (Pardo and Patterson 2013, p. 198). Of course the social function of criminal courts is different from that of civil courts, but this should not prevent us from using neuroscience as a support tool in the various areas of civil law as well, nor is there a conceptual barrier preventing us from making such a use of it (Alces 2018). And indeed this has started to happen in diagnosing consciousness disorders, such as vegetative states, aphasia, or minimally conscious states (Farisco 2013), as well as in aiding end-of-life decisions (Fuselli 2014, p. 15; Ragone and Vimercati 2017). Generally speaking, any area of private law is amenable to neuroscientific study (Desmoulin-Canselier 2017; Tafaro 2017). The next step will presumably be to use neuroscience to understand what elements are involved in a person's capacity to make contracts or in determining mistake and good faith, due diligence, or the degree of guilt or responsibility, all issues in private or civil law.

Neuroscientific evidence, organized along the *continuum* that lays out the probabilistic distribution of reality, clashes with the conceptual structure of the law, which instead is organized on the basis of *categorial* dichotomies. Under these dichotomies, for example, one is either guilty or innocent or is responsible or not responsible (Sapolsky 2006, p. 230). The interaction between these two worlds brings out the "disconnect between the philosophy (and science) of the 'theoretical' mind and pragmatic and commonsense positions," where models based on the latter can only prove "primitive and scarcely justifiable in light of recent experimental discoveries" (Sartori et al. 2011, pp. 136–137). The consequences may be relevant. Precisely on the basis of a neuroscientific paradigm, a new foundation for criminal law could mean that we should strip away from the law the last vestiges of the retributive principle, redesigning the legal system as one of "pure social defense" (Sartori et al. 2011, p. 163). This contradicts the innate tendency to afflictively punish socially despicable facts.

Neuroscientific evidence has gradually been accepted as an object of expert testimony, depending on the standard of admissibility provided in each system. The case law has developed a range of doctrines in this regard, but they all invariably refer to one or more external criteria of admissibility, such as whether the scientific laws in question are falsifiable, whether the research has been published through a peer-review process, whether the error rate is acceptable according to consolidated protocols, and how widely accepted the research is in the relevant scientific community, similarly to what in U.S. case law has been established in *Daubert v. Merrel Dow* (Garland 2004a; Shafi 2009).

Even in countries where these criteria seem more stringent, it is still the judge who ultimately acts as *peritus peritorum*, in full control of the decision as to whether the evidence is scientific. The neuroscientific technology now in widest is neuroimaging, providing data on the structure and functionality of the brain (Goodenough and Tucker 2010; Greely 2004; Oliverio 2012; Pineda 2009; Santosuosso 2011; Shafi 2009; Tancredi 2004; Genevès 2016). Neuroimaging is generally treated as a single technology in the legal literature, but in fact the name covers a range of very different techniques, and it is necessary to analyze them systematically.

The first difference is that between *structural* and *functional* neuroimaging. Structural neuroimaging is designed to investigate the brain's structure in a static way. The two main techniques are *magnetic resonance imaging* (MRI) and *computed tomography scan* (CT). MRIs are noninvasive, and because of the type of resolution they offer, they are considered a scientifically reliable measure in diagnosing brain injuries. However, there are subjective elements in the measurement, depending on the clinical protocols that are applied, which in some cases have presented a potential problem for the use of data in court. Moreover, as with all clinical evidence, even univocal results do not always count as sufficient evidence: while brain injuries can impair judgment and memory, they do not necessarily make one more prone to crime.

It has been clearly acknowledged that prefrontal cortex damage is linked to behavioral disorders, for the cortex plays a role in decision-making cognitive functions. This is the brain area that only matures later in a person's life, in some cases not until adulthood, and the discovery of its relevance has been among the grounds for making it unconstitutional to sentence minors to death (Glannon 2014; Messina 2015, p. 259; Pizzetti 2012, p. 33). And yet the causal link to criminal behavior is far from certain. There is no conclusive evidence that these injuries to the prefrontal cortex, even serious ones, can be used to predict behavior, if only because many people who have sustained such damage do not go on to commit any kind of crime (Sapolsky 2006, p. 239). Like MRIs, CTs are noninvasive and offer high spatial resolution, but they also offer high temporal resolution, making it possible to follow the functioning of the organ. However, it is the MRI that stands as the main tool for assessing the damage inflicted by trauma or brain disease. These two methods are widely established in clinical practice, but they are not decisive in assessing mental abilities, even if the courts have sometimes admitted such evidence in determining criminal capacity.

In one of the most famous court cases where brain imaging was used—the 1982 trial of Ronald Reagan's shooter (*United States v. Hinckley*)—the defense attorneys asserted the insanity defense, arguing that the defendant was mentally impaired (and so should not be held responsible), and to this end they introduced as evidence a CT scan showing atrophy of the brain. The expert witness appointed for the psychiatric evaluation claimed that the type of atrophy revealed in the CT scan was associated with schizophrenia. The testimony of a radiologist, however, excluded that those brain abnormalities had any necessary causal effect on one's behavior or mental capacity. The jury accepted the defense's argument and acquitted Mr. Hinckley. To be sure, the neurological expertise was only one of the elements introduced in raising the insanity defense, but according to some commentators it was likely a decisive element (Khoshbin and Khoshbin 2007, p. 184). Several years later, a teacher was acquitted of pedophilia because it was shown that the frontal cortex had been inhibited in its control functions by an orbitofrontal tumor that pressed on the area. Once the tumor was removed (twice), the teacher ceased his behavior. He claimed that, while he realized that the criminal acts for which he was being tried were reprehensible, he was unable to prevent himself from carrying them out (Burns and Swerdlow 2003).

A further brain imaging technique is functional neuroimaging, which is used to measure metabolic activity and brain processes in action. Functional techniques measure neochemical processes to develop an individual cognitive profile (Baskin et al. 2007, p. 250) arrived at by analyzing the inflow and outflow of blood in different areas of the brain. One of these techniques, *single photon emission computed tomography* (SPECT), is used to diagnose various neurological and psychiatric disorders, including dementia, epilepsy, depression and schizophrenia.

The SPECT functional technique is even less reliable than structural ones, especially because of its lower spatial resolution, and so its forensic use is still very limited. Another functional technique—more accurate than SPECT but with a similar spatial resolution—is *positron emission tomography* (PET), which measures metabolic activities, and in particular blood flow and volume, by injecting a radioactive contrast liquid into the circulatory system (Pettit 2007, p. 320). PET results are influenced in part by a subject's age and clinical history, as well as by other contingent conditions, and depend in any case on the control group used.

Finally, *functional magnetic resonance imaging* (fMRI or RMF) measures local brain activity based on the flow of blood and the use of oxygen in circumscribed regions of the brain (Appelbaum 2007, p. 461). fMRI detects hypoactivation and hyperactivation, signaling irregularity in those regions. This technique, too, is noninvasive, and scientifically it is considered superior to the other functional techniques.

An fMRI produces very accurate anatomical images, and at the same time it provides a neural evaluation of cognitive and behavioral activities (Metting et al. 2007, p. 705). The potential is for fMRIs to become a powerful tool for psychological profiling and for assessing the reliability of witnesses (Kittay 2007). As concerns the evaluation of witnesses, however, there is not yet a scientific consensus on the neurological basis of deception, since we do not yet have sufficient experimental data on error rates. There are also external variables, such as the level of attention or contingent emotions, that may end up affecting the reliability of results. For this reason, the ethical concerns and constitutional implications are similar to those that come up in connection with use of the polygraph or of other lie detection devices (Arrigo 2007; Beecher-Monas and Garcia-Rill 2015). Then, too, the main use of fMRI is still in research rather than in diagnosis. This is because some of its considerable limitations, chief among which are the poor temporal resolution of the image it produces; the relative nature of its results, in that they need to be defined relative to a reference group; and the nonnegligible possibility that the data may be influenced by anatomical differences among subjects or even by any imperceptible movements they may make.

As much as fMRI is sophisticated and accurate in several respects, it is often considered unreliable in court. One of the most interesting cases involving the use of fMRI came up in 2005 in a dispute between an association of video game developers and the State of Illinois (*Entertainment Software Association v. Blagojevich*). Illinois had made it illegal to advertise violent video games to minors, on the theory that exposure to this kind of entertainment would remove the inhibitions that prevent young players from enacting the violence they see on the screen. The video game developers therefore challenged the ban in court, and the state, in its defense, produced an fMRI study that seemed to confirm that violent media content (and so also video games) did in fact encourage the violent behavior the ban was intended to prevent. The plaintiffs criticized the study on two grounds—namely, that the subjects it analyzed were only *simulating* the use a violent video game, without actually playing it, and that it failed to show a causal link between reduced frontal-lobe activity and aggressive or violent behavior (Feigenson 2006). The judge agreed that the study was unreliable as evidence for the state's claim (Shafi 2009, p. 37).

Despite the concerns just noted about the reliability of neuroimaging, this technology is increasingly being presented as legal evidence (de Cataldo Neuburger 2009), even if occasionally only *ad abundantiam* (Garland 2004b, p. 20). The particular appeal exercised by studies on the brain, and the self-explanatory and suggestive force of the images so produced, have in some cases proved very effective in influencing lay judges and juries, and perhaps even professional judges. Bertolino (2013, p. 161) reports studies from which it emerged that verdicts of not guilty by reason of insanity in the case of personality disorders increase significantly (from 12 to 42%) when neuroimaging evidence is provided that the accused has suffered a cranial trauma and damage to the frontal lobes. In addition, other studies have found that the use of neuroimaging becomes more effective when the evidence is accompanied by a historical reconstruction of the mental disorder, and by an assessment of the accused in person, which makes it possible to see firsthand what he or she is like, so as to intuitively form opinion of one's own about this person (Messina 2015, p. 260).

In the case cited by Raine (2014, p. 135), the image of a cerebral cyst, used to claim that the brain of the accused had been damaged to the point of inhibiting self-control mechanisms, resulted in a conviction that was reduced from murder to manslaughter. In another case, the state of neurophysiological alteration caused by premenstrual syndrome was used, this time unsuccessfully, to argue for a transitory mental incapacity (Rose 2007, p. 230).

In a case reported by Pizzetti (2012, p. 34), a defendant was found guilty of murder on the basis of an electro-oscillation. From the exam, there emerged a cerebral familiarity with some details of the crime that were unknown to the public, having been reconstructed by the investigators and known only to them. Despite the neuroscientists' own skepticism about the evidentiary value of neuroimaging as a tool by which to establish behavioral traits or patterns, there is a significant proportion of cases where different jurisdictions have accepted this technology as reliable evidence (Shafi 2009, p. 34). According to one study (Feigenson 2006), this proportion was over 80%.

Although structural investigative techniques are more well-established and scientifically sound than functional ones, it must be considered that, even in regard to the former, the judicial reconstruction of reality is necessarily more stringent than the clinical reconstruction. The clinical limitations of SPECT and PET, for example, may lead to considerable discrepancies in the interpretation of outcomes, and in the judicial system, where the prohibition on discriminations between subjects and the principle of equality are paramount, they are likely to be a source of judicial error. Courts should be more than cautious when images are obtained with clinical protocols that are not well established. Images of the brain's structure or activity are taken some time after the crucial event, so there is no way to be certain about what was happening at the legally relevant time (Glannon 2014, pp. 157–158). Forecasts based on neuroimaging are made by reference to statistical populations, and inferences based on the relative biomarkers should be sufficiently robust even at the individual level. Otherwise they cannot even partly motivate a judicial decision (ibid., p. 164).

In connection with this specific problem, Schauer (2010, p. 102) claims that it may be difficult to achieve a balanced relationship between neuroscience and law. These two logical systems, and their orders of values, overlap only in part: what can be satisfactory for neuroscience may be unsatisfactory for the law, and vice versa. It has also been argued, however, that forecasts of individual violent behavior cannot be considered inadmissible, since the same level of reliability is normally accepted for standard clinical assessment (Lippert-Rasmussen 2014, p. 132). These tools do not lead to measures permanently limiting freedom, but they rather serve as elements on which basis to motivate provisional measures that are only slightly afflictive (usually light security measures or targeted searches). Personality disorders have been recognized by the Joint Session of the Italian Court of Cassation as part of the list of mental incapacities specified in Articles 88 and 89 of the Italian Criminal Code (Court of Cassation, May 5, 2011, CED 250067). After all, except for the case cited by Khoshbin and Khoshbin (2007, p. 184), under no circumstances was the determination of a mental illness entirely based on neuroscientific evidence. We are therefore not at risk of neuroscientism, while neuroscience could contribute to increasing the scientificity of decision-making, especially from a methodological point of view (Bertolino 2013, pp. 159–160).

Neuroevidence is in a way an extension of psychological evidence: it may be useful, for instance, in assessing the severity of punishment (Amato 2013, p. 37). In one case, the Trieste Appeals Court of Assizes accepted that an individual's genetic characteristics can act as triggers of aggressive behavior, and partly on this basis it reduced the sentence from ten years in prison to nine (Capraro 2013; De Felice 2010, 2014; Pizzetti 2012; Sammicheli and Sartori 2010). In making the case for the reduced sentence, the defense had presented fRMI results showing that when the accused was administered a stop-signal task, he had below-average control abilities and greater impulsivity, and also below average were his attention span and cognitive alertness (Sammicheli and Sartori 2010, p. 210). The subject was in fact inclined to act aggressively and impulsively under stress (Capraro 2013, p. 174).

When is it that an individual behavioral trait of organic origin is grounds for a finding of mental impairment—meaning that the accused lacks full capacity and therefore cannot be found fully responsible, justifying a more lenient judgment against him or her—and when is it that the same trait points to a subjective inclination to engage in antisocial behavior that instead legitimizes a harsher sentence? The conceptual categories of Italian criminal law may not help when it comes to addressing this question, but the practical effects, as we have seen, are in most cases limited. The decision of the Trieste Appeals Court of Assizes was widely criticized as based not on an assessment of the defendant's psychology but on his marginalized position as a noncitizen and on his readiness to snap in response to religious slurs (Aleo 2010, p. 240). It was also argued that if a harsher punishment should be imposed, it should fall on those whose behavior was more aggressive, regardless of provocation, in the interests of crime prevention, and that if there is a societal interest in protecting those who are marginalized, that aim needs to be pursued by recourse to security measures, that is, by committing the offender to psychiatric hospital by reason of insanity or diminished mental capacity (Aleo 2010, pp. 244–245).

In yet another Italian case, the neuroscientific evidence was used by the Court of Cremona to ground the part of its decision that concerned the reliability of the claims made by a victim of sexual harassment. Based on this evidence, the victim's recollections were deemed "proper and natural," and this contributed to the defendant's conviction (De Felice 2014, p. 96). In this case the court relied on the so-called implicit association test (IAT), measuring how strongly concepts are associated in a person's mind by looking at the person's response latency (Capraro 2013, p. 179).

At a pretrial hearing in Como, in 2011, neuroscientific test results were presented as part of the evidence in reconstructing the overall picture of the individual's psychological situation: the tests revealed a morphological anomaly in the defendant's frontal lobes (Capraro 2013, p. 175), and the court found that this result was admissible as evidence. The defendant had attempted to kill his mother and then killed his sister (Pizzetti 2012, pp. 121–122). The court took the view that, although these tests could not replace a behavioral and clinical assessment of capacity, they ought to complement them, and so when two clinical interviews were presented that had been performed without administering psychodiagnostic tests, the court held that they were an insufficient basis on which to decide the factual issue in one way or another. Thus the court placed greater trust in objective imaging measurements and molecular genetics. Such neuroscientific insights reduce diagnostic variability and discretionary responses (Messina 2015, p. 249). The previously mentioned 2005 decision of the Italian Court of Cassation envisions an integrated model for ascribing criminal responsibility, in which each expert opinion must be placed within a broader evidentiary framework, and in which the "organicistic paradigm" for establishing mental illness is superseded in favor of an approach that is not just clinical but also psychological and sociological (Messina 2015, pp. 255–256).

It is becoming increasingly difficult to defend the view that the role of neural factors is only superficial and is not reflected in practice (Terracina 2013, p. 166). At the same time, however, the potentialities of neuroevidence should not lead us to ignore the critical issues raised by the forensic use of neuroscience (Sartori et al. 2011, p. 156). There is a risk that the use of genetic and neurological investigation techniques aimed at determining vulnerabilities or levels of risk should create a "biological underclass" of people without education, without insurance coverage, and unlikely to be attractive in the labor market (Tancredi 2004, p. 112). The fundamental ethical dilemma concerns the risk that once neuroscientific data is acquired in the interests of justice or for the care and protection of people, it may end up being used as a basis of discrimination. In fact, the social and ethical implications of new technologies become clear only when they are fully developed and in widespread use (Garland 2004a, p. 8). This is the concern expressed by D'Avack (2013, pp. 24–25), in response to which we should be able to say that these practices are sufficiently safe if they strictly follow scientific protocol and include both a risk assessment in view of expected benefits and the subject's free and informed consent.

Forensic neuroscience makes it possible to obtain evidence consistently with criminal procedure: it could provide the evidence needed to establish the facts on which basis to determine whether a crime can be attributed to someone (imputability), whether this person is to be found guilty of the crime (culpability), and, if so, whether its punishment should consist of a prison sentence or a security measure (Pizzetti 2012, p. 52). This is not to say that neuroscientific tests cannot be used in such a way as to undermine the rights of the accused. Thus, for example, they could be uncritically used as the main or sole basis for a conviction rather than as a source of greater knowledge in seeking to establish the *real nature* of the defendant (Pizzetti 2012, pp. 116–117).

The danger is always lurking because, while it may not be obvious, the images produced using scanning techniques are not in fact produced by the scanner but by human decisions about the way the data must be treated and highlighted, and about the standards used in making these decisions (Jones et al. 2013, p. 734). It may be unnecessary to point this out if the judge appreciates what the correct use of scientific evidence is. But any technological instrument or new technique is a source of problems in the hands of an intellectually lazy judge or lawyer, especially considering that its limitations may not have been fully understood.

4.3 Imitation, Mirroring, Empathy

Neuroscientific investigations have contributed to one of the most remarkable discoveries in the understanding of social behavior. I am referring to *mirror neurons*, which shed light on the neuronal bases of the instinctive behaviors of imitation and empathy. The relevance of basic emotions in legal behavior has been extensively treated in another part of this work (Sect. 1.4). The research on mirror neurons seems to provide an organic foundation for a substantial part of this emotional component, or at least it more clearly illustrates its biological substratum.

Research on mirror neurons began about twenty years ago with a group of neuroscientists and neurophysiologists of the University of Parma, and it has since been constantly updated by the use of more advanced analytical tools (Ferrari and Gallese 2007; Rizzolatti and Craighero 2004). The discovery of mirror neurons has attracted a body of other studies that have made it possible to build a vast knowledge base about their implications, providing a starting point for new experiments (Bråten 2007; Hari 2007; Iacoboni 2009; Oberman and Ramachandran 2009). As we will see further on, there have been several criticisms, which in many cases have fundamentally noted how the core of the mirror-neuron theory has sometimes been exceedingly stretched to cover and explain phenomena that were out of its reach.

The initial experiments that led to this discovery were done on primates. Quite accidentally, it was discovered that certain areas in their brain would "fire" when the individual perceived another's action. It was thus hypothesized that the observer possessed an internal repertoire of actions through which mirror neurons allowed him to form an implicit and immediate (and therefore empathic) understanding of what he was observing. The mechanism, also present in humans and any living being with a sufficiently evolved brain, has provided neurophysiological confirmation that the mind is simulative: neurobiological processes underlie the processing of simulations in different areas of cognition, including social cognition.

However, the terminology of *mirror neurons* that has come out of this research line was likely misleading, because we do not actually have single neurons that functionally belong to an area dedicated to mirroring (or at least, their existence has not yet been proven): there rather appears to be a neuronal system dedicated to mirroring that works through its neurons in the brain of evolved beings (Rizzolatti and Craighero 2004, p. 174).

In some brain areas neurons were identified that were activated by "finalized motorial acts," and were therefore capable of understanding an action's potentiality for action, regardless of how it is in fact carried out (Rizzolatti and Sinigaglia 2006, p. 2). On the simple basis of the associated intention (Iacoboni 2009, p. 130), the individual recognized the meaning of the action and understood what others were really doing, beyond the sensory-pictorial representation of it (Rizzolatti and Sinigaglia 2006, pp. 3–4). There would therefore be no distinct areas for perceiving and interpreting the visual stimulus. Most of the neurons contained in specific areas would be dedicated to the perception of intentional acts. Transcranial magnetic stimulation studies have established that in humans, these areas are larger, and that mirror neurons can also encode the temporal dimension of the single acts that make up a movement.

Mirror neurons play an implicit understanding function, of pragmatic and nonreflexive origin, disconnected from any specific sensory modality, but bound to the potential actions inscribed in the vocabulary of acts that in each individual regulates and controls the execution of movements (ibid., p. 104). Their primary role is therefore linked to understanding the meaning of entire chains of acts. The current hypothesis is that this is an adaptive mechanism by which to anticipate risky situations, that is, to understand what they are ahead of time and to accordingly act to prevent them.

Similarly to the immune system, which does not have a precise spatial localization but maintains a unique function and a response pattern in various parts of the body, Oberman and Ramachandran hypothesize that the mirror-neuron system serves to "connect our own representations with those of others across multiple domains, and more generally to map one dimension onto another in order to abstract what is common to them" (Oberman and Ramachandran 2009, p. 51). The mirror-neuron system is seen as the "neural foundation on which basis human beings can make mental simulations, considering that it is active in many social activities: imitating others, understanding their mind, feeling empathy for them, and intentionally being in sympathy with them" (Anolli and Mantovani 2012, p. 116). Imitation, in particular, is a fundamental activity of the human species: it forms the basis of culture and learning, and it is how we form personal theories of the mind, that is, how we come to understand the mental states of others. Imitative behavior is understood to elicit appreciation in those who are imitated. Thus, for example, experimental research in psychology has shown that tips at a restaurant significantly increase simply by repeating a customer's words when taking an order (van Baaren et al. 2003).

Imitation is a form of social glue, because it unwittingly leads to better relationships, mutual appreciation, and simpler interactions. Priming dynamics also demonstrate the unconscious effect of relational dynamics which are apparently irrelevant to the substance of the relationship, but which actually have a noticeable impact on intellectual performance and interpersonal behaviors (Franks 2010, p. 174).

Imitation has reasons more profound than mere opportunism or fear, as we are led to believe: it automatically kicks in much more often than we think (Franks 2010, p. 41). We cannot empathize with other and understand their reasons if we cannot imitate them. A century before the concept of *social empathy* was developed, the Icelandic sociologist Finnbogason had anticipated it in his theory of *sympathetic understanding*, which identified the phenomena that would later be described by neurological research on mirror neurons:

> En imitant l'attitude d'un autre à l'égard d'un objet donné, on pourrait au moins approximativement acquérir sa façon de voir cet objet. Après nous être fait un idée claire de la nature de l'accommodation, et avoir pris égard aux expériences de gens qui s'appliquent à mimer les autres, il nous semblé être de plus an plus en droit d'affirmer que nous pouvons, en réalité, comprendre l'individualité des autres au même degré qu'une imitation volontaire ou involontaire de leur manier d'être et d'agir fait passer leur individualité dans notre âme et notre corps, bref, nous accommode comme eux. Mais, une fois qu'ils sont passés en nous de cette façon, nous pouvons nous rendre compte de la manière dont ils penseraient, sentiraient et agiraient dans des situations variées, à condition qu'ils agissent d'accord avec le côté de leur nature qui est passé en nous. (Finnbogason 1913, pp. 229–230)

The simulation "embedded" in mirror neurons would enable us to understand not only an action but also the intentions behind it (Iacoboni 2009, p. 132; Lyons 2009, p. 127). In a sense, this can be regarded as an elemental form of "mind reading" (Ferrari and Gallese 2007, p. 85). The mechanism of mirror neurons could also be taken as confirmation of Mead's thesis of gestural conversation as an expression of innate sociality (Baggio 2015, p. 77).

The sociological hypothesis we can derive from the theory of mirror neurons is highly counterintuitive. The task of understanding the motivations of others is not an intentional deliberative process but is rather an instinctive-motorial activity: we involuntarily identify with others to the point of also reflecting, for example, their competitiveness. This happens to all spectators of sports: they find themselves moving their own body as if to help the basketball into the hoop, or jumping a hurdle together with the horse and jockey. This mirroring performed by spectators at a sports event has led Rizzolatti and Sinigaglia (2006, p. 147) to speculate that this is a "motorial release" dependent on an intense emotional participation through which control mechanisms are attenuated: when we participate and identify in this way with action unfolding before us, we are prompted to make movements or do acts that would otherwise remain in the potential state.

Some have suggested that mirror neurons do for psychosociology what DNA has done for biology, that is, they provide a theory which unifies several paradigms. A definitive confirmation that there is a class of neurons that serves this function in humans would link up apparently unrelated phenomena (imitation, our formation of theories of the mind, empathy, and language). A neural system of this kind would serve as a foundation of social competence, and would offer the first neuroscientific evidence of a secularized morality, evolutionarily constructed through neurobiological mechanisms that enable us to understand the intentions of others (Iacoboni 2009, p. 132).

The possible functions where the mirror-neuron system could have a role are summarized by Oberman and Ramachandran (2009, p. 42ff.) under four headings: (1) synchronization of behaviors, (2) imitation, (3) language (if we subscribe to the idea that language at its core is not made of acoustic signals but is rather made of "phonetic gestures"), and (4) empathy and the ability to form theories of the mind. The fact that the pronunciation of certain words (like *bouba*) is almost invariably associated with a spherical shape, while other sounds (like *kiki*) are instinctively associated with a pointed shape, could be explained by postulating that a zone of multi-sensorial neurons, potentially mirror neurons, is tasked with mapping certain inflections by associating them with a visual stimulus (Oberman and Ramachandran 2009, p. 46). However, one can never be too careful when venturing into such all-encompassing explanations. There is no lack of compelling and even radical criticisms of the "myth" of mirror neurons (Hickok 2009, 2015). Mirror neurons cannot explain a range of phenomena so wide as to include everything from the tendency toward obesity to the aptitude for appreciating music (Hickok 2015, pp. 32–33).

After pointing out all the conceptual and theoretical problems of the mirror-neuron theory, and the inconsistencies of its all-explanatory use, Hickok (2015, p. 202) proposes that we go back to the starting point, without venturing beyond the attempt to understand our choice of movements on the basis of interpersonal relationships. Even if mirror neurons simply reveal a new form of conditioned reflex that remains in the neuronal state, this discovery is by no means less significant. As has also been observed by those who have discovered mirror neurons, "sharing the emotional state of others on a visceral-motorial level is not the same as trying empathetic involvement with them" (Rizzolatti and Sinigaglia 2006, p. 181).

What is beyond doubt is that basic social attitudes like imitation and empathy have a neural substrate, and they can also be analyzed from this further point of view. Brain scanning has helped to reframe in original ways the classic problem of intersubjectivity (Lyons 2009, p. 122), or how people can understand and share the mental states and emotions of others. Mirror neurons offer an explanation based on simulation: we can understand the mental states of others by literally putting ourselves in their shoes, without needing any further logico-cognitive steps. Imitating and mirroring others are experienced as straightforward processes, but in reality these activities require considerable interpretive effort, which we are not aware of, because we have an instinctual system specifically dedicated to it (Lyons 2009, p. 79).

Human intersubjectivity develops in four phases: from the second month of life, mirroring, imitation, and other types of primary emotional responses of contagion are overcome by the "first signs of reciprocation (primary intersubjectivity), joint attention to objects (secondary intersubjectivity), the emergence of values that are jointly represented and negotiated with others (tertiary intersubjectivity), and eventually the development of an ethical stance accompanying theories of mind by 4 years of age" (Rochat and Passos-Ferreira 2009, p. 191). The most updated developmental psychology—which breaks with the Freudian and Piagetian tradition, anchored to a vision of the immature subject as fundamentally self-centered—is advancing the hypothesis that the newborn has an innate capacity for interaction and "alter-centered" participation. Relational maturity at the neuronal level is already much more advanced from the first moments of life than previously thought, making it possible for the newborn to respond and react immediately to stimuli (Ferrari and Gallese 2007, p. 73). Imitation and mirroring are, in effect, a tit-for-tat system, although in order to develop a properly social attitude, we still must be supported by open reciprocation and negotiation (Rochat and Passos-Ferreira 2009, p. 192).

Rochat and Passos-Ferreira (2009, p. 199) make the example of the handshake. This represents either the culmination or the beginning of a negotiation process, and it is sealed with a mirror-like gesture. A particularly interesting and counterintuitive example of the control the higher functions instinctively exert on a complex motor impulse is *yawning* (Norscia and Palagi 2011). The ease with which the impulse to yawn imposes itself on our cognitive rationality—not only by observing someone yawn but also by hearing it mentioned—constitutes an always surprising confirmation of the impossibility of describing our impulses as rational and voluntary (Hari 2007, p. 93).

With some rare exceptions (Gregory 1983; Kendon 1970), imitation, camouflage, and synchronization are among the basic processes that are ignored in the conceptual framework of much sociological theory. Yet copying allows us to innovate and build intersubjectivity (Rochat and Passos-Ferreira 2009, p. 197). Contagion forms the basis of the social processes that lead to the emergence of a cultural or intellectual fashion. In the neuroscientific literature, emotional contagion is assimilated to imitation: it consists in experiencing (without reason) the same things that someone else experiences. In empirical research, reference may obviously be made to the basic emotion of empathy. When we observe emotions of pain, disgust, and fear in other people, our brain activates the same nervous circuits: the insula, the anterior cingulate cortex, the amygdala, and the superior temporal sulcus (Anolli and Mantovani 2012, p. 130).

Thanks to new findings in neuroscience, empathy has attracted renewed attention in philosophy and the social sciences (Boella 2006; Bråten 2007; Calvet Christian and Alm 2012; Gallacher 2011). Empathy, which had been obscured by the ethics of sympathy or compassion, enables us to "realize that someone else is a subject of experience just as we are: we experience feelings and emotions and perform volitional and cognitive acts" (Boella 2006, p. xii). The feeling of empathy is the outcome of a process of intentional attunement that superimposes the intentions of the subject being observed over those of the observer: in this way it provokes an impetus of familiarization, and it ultimately transforms *the other as an object* into *the other as another self* (Ferrari and Gallese 2007, p. 86).

In order to move beyond a purely ethical dimension of empathy, we need a broader understanding of it. Empathy is not necessarily an instrument with which do good, and in fact it may even be used as a tool of deception or to inflict suffering. There is an empathy of the predator (Tisseron 2010, p. 17): the sadist who empathizes with his victim can understand how to inflict even greater pain (Boella 2006, p. 116), and deception uses empathy to advantage as a tool by which to win the trust of those who are being deceived. These examples illustrate that empathy is ethically neutral (Gallacher 2011, p. 4). As Bloom (2016) convincingly argues, moral decision-making should be based on "rational compassion" rather than on empathy.

Even the emotions of others are transformed into representations of own's emotions, with the collaboration of the areas involved in the corresponding facial expressions. When we simulate the brain activity of other individuals through our own brain activity, we assume that others are similar to us (Keysers and Gazzola 2009, p. 24). If these areas sustain damage, the affected individual also finds it difficult to recognize the facial expressions of others; when they are particularly active, by contrast, they determine the greatest degree of empathy, and not coincidentally, this is negatively correlated with autism (Keysers and Gazzola 2009, p. 23).

Human sociality may sprout by way of imitation, but only through reciprocation does it bloom (Rochat and Passos-Ferreira 2009, p. 207). The mechanisms of imitation, mirroring, and empathy cannot in themselves be the lens through which to interpret any social system or subsystem, including the legal system. In his study on Mead, Baggio (2015, p. 65) observes that imitation alone, without cooperation, remains an end in itself. However, when these mechanisms are regarded as extemporaneous phenomena beyond the scope of sociological analysis, a serious flaw vitiates its analysis of behavior. Because imitation and empathy are hardwired, they are also relevant in explaining legal attitudes and behavior. Blumenthal (2005, p. 23) experimentally observes, for example, that humor greatly affects greatly all kinds of moral and legal judgments. In the study he found that when jurors in criminal trials felt annoyed, their judgment tended to result in guilty verdicts rather than in acquittals, especially when their annoyance was with the behavior of the defense rather than that of the prosecution.

Gallacher (2011) and Nussbaum (2006), among others, argue that our ability to empathize (or, rather, our ability not to suppress our instinctive tendency toward empathy) should be cultivated as a useful professional skill even among lawyers. Lawyers who remain cold and detached from their clients—acting as rational analyzers of legal rules, without considering what it is to personally experience their effects—will not be able to effectively advance their clients' cause. In fact, a lawyer so described wouldn't even be able to grasp the deeper motivations of the counterparty, and so wouldn't see the risks the client runs. As Nussbaum observes (2006, p. 325), the training that lawyers are put through is hyper-rationalizing: she argues that this takes empathy out of the picture and thus makes it more difficult to relate to the parties in conflict, and hence to understand the conflict itself and the law that applies to it. Using the rediscovery of empathy, Di Giovine (2009, pp. 176–177) supports the need for a return to the "factual" even in criminal law, and lists a series of bioethical issues (such as organ donation and medically assisted procreation) where we could gain deeper insights in light of the renewed awareness of emotional arousal.

Even in business law it will not be possible to benefit from a knowledge of the underlying facts without realistically grasping their "human meaning," for a business lawyer must always be able to understand what reactions a business strategy will draw from consumers and workers and in public opinion (Gallacher 2011,

p. 36). In the following sections, I will deal with a further series of phenomena that are hardwired into our brains, and we will see how they directly impact law-informed behavior.

4.4 Decisions, Heuristics, and Biases

The idea of limited rationality introduced by Herbert Simon was subsequently investigated by other sociologists and economists (Cicourel 1974; Collins 1975; Schütz 1967). In his more ambitious theoretical essays, Merton (1968, p. 474ff.) develops a theory of behavior that is centered on reference groups and adheres to an idea that was then developing in the social sciences: that of the self-fulfilling prophecy as a pervasive social mechanism. This is the idea of a situation that leads to real consequences simply by virtue of its being *framed* or *defined* that way.

These mechanisms—such as compulsive conformism or anticonformism, acute inertia, or the obsession with novelty—are considered by Elster (1985, p. 22) as "non-autonomous preferences." Heuristics that are not strictly rational, or that respond to a different rationality, were subsequently fully understood in cognitive research, primarily through the research line that Kahneman and Tversky (1974, 2000a) initiated in the 1970s. The studies done in this field, often in cognitive psychology but in fact multidisciplinary, highlighted the inadequacy of the traditional models used in describing and understanding the way humans behave in making decisions (Rumiati and Lotto 2011, p. 193).

In Sect. 1.3, we discussed the concept of rationality and its evolution with specific reference to law. Law is in large part aimed at inducing a change in human behavior, and therefore presupposes a conception of how behavior is encouraged or deterred. With the exception of classical economic theories, however, it is rare for this conception to be made explicit. There are various models that the law either looks to or presupposes in predicting behavior: they may be based on experience, observation, intuition, or the moral sense, and the outcome is unsystematic, with lawyers haphazardly picking up different ideas from various disciplines, while still systematically ignoring the cues that come from the biological and life sciences (Jones and Brosnan 2008, p. 1954).

The unsatisfactory way in which law relates to human nature is essentially the polemic goal of Jerome Frank's realism. Frank (1949, pp. 145–146) goes so far as to claim that law faculties should also teach "psychology applied to law in all its phases." As much as the law may be conceived as a rational instrument, Frank unmasks its unconscious biases and prejudices (especially in juries but also in judges), against women, minorities, or people with physical defects (ibid., p. xi), and he points out some misconceptions, such as the myth that the law is uncertain not because the facts themselves are uncertain but because we are uncertain about the rules, or the myth that errors of judgment can be rectified in the higher courts.

Frank's realism is particularly modern. He argued that law consists not of rules but of decisions, in that rules are just "psychological pulleys," "mental bridges or ladders" (Frank 1949, p. 167) relying on which judges can select the relevant facts that will make their decisions logical. The myth of legal certainty, and of its quest, is ascribed by Frank to an unconscious Freudian tendency, that is, to the desire for a strict, infallible, controlling father figure (ibid., pp. 19, 248), or to the impulse of religious spirituality (ibid., p. 196ff.), whose ultimate goal would be similar: to grasp a divinity that leaves no uncertainties.

In truth, in subsequent works Frank (ibid., p. 263) broadened the list of possible explanations for the myth of legal certainty: he included the aesthetic desire for symmetry, professional habit, the protection of economic interests, security instinct, the yearning for peace, imitation of custom and devotion to it, inertia, laziness, stupidity, mindset, the magic of ritual linguistic formulas, and language and word magic. Frank (ibid., p. 252) wanted modern law to be freed from the authority of a paternal or divine figure and become "avowedly pragmatic."

The desire for certainty, whether or not linked to one's relationship with the father figure, is correctly perceived by Frank as a foundation not only of the legal sentiment but also of other social feelings. Among the drivers of choice and behavior is the desire to eliminate inconsistencies and "cognitive dissonances." Individuals aim at self-consistency and try to suppress dissonance by actively steering clear of situations and knowledge that would reinforce it (Festinger 1957, p. 3; Wexler 2008, p. 155ff.). One of the first discoveries made by the cognitive sciences, when the system of human judgment was subjected to in-depth analysis, was that we are naturally ill-equipped to reason inferentially on a statistical basis (Gintis 2007, p. 12). In our search for the truth, we use *a contrario* reasoning: through a Bayesian process (Bona and Rumiati 2013, p. 110ff.), coupled with sequential decision-making (Magrin 1995, p. 178), we assess the probabilities that an event will take place on the basis of our observation of events that have taken place in the past (Goldman 2012, p. 165).

To use Simon's categories, the decision-maker's guiding model for economizing cognitive resources is not that of the *maximizer* but that of the *satisficer* (a portmanteau of *satisfy* and *suffice*). Considering that we must rely on intuitions, habits, and an incomplete observation of reality, the necessary limits of rationality lead us to content ourselves with results that we regard as no more than satisfactory (Jones et al. 2001, p. 1151). Kahneman and Tversky identify some basic guiding criteria, that is, three heuristics based on natural evaluations that people automatically use to estimate the probability and frequency of an event or the magnitude of a value: the availability heuristic, the representativeness heuristic, and the adjustment-and-anchoring heuristic (Kahneman and Tversky 1974, p. 1124ff.). "Heuristic principles reduce the complex tasks of assessing probabilities and predicting values to simpler judgmental operations. In general, these heuristics are quite useful, but sometimes they lead to severe and systematic errors" (ibid., p. 1124). These systematic errors are called cognitive biases.

Cognitive biases are not a defect of rationality but rather the result of adapting to a complex world (Haselton and Nettle 2006, p. 59). With the availability heuristic, we judge the frequency or probability of an event on the basis of the ease with which we recall that event. With the representativeness heuristic, we judge the frequency or the probability of an event on the basis of its similarity to other events or classes of events and to their probability or frequency. With the adjustment-and-anchoring heuristic, we make estimates starting from initial anchor values, moving away from them only through marginal adjustments that frequently prove to be insufficient.

Because the human mind is not equipped to use theories of probability, the economist's classic metric of expected utility cannot be a guide for predicting individual choices. We are not instinctively capable of understanding the distribution of *default* probabilities (the *base rate*). The fundamental contribution this research line has made to an understanding of decision-making lies in prospect theory. By *prospect* is meant a cognitive, objective, and external point of reference that provides the *natural* term of comparison taken as a starting point for our evaluation. In behavioral terms, we do not know what we want until we can place it in context and evaluate its relative advantage over an alternative, thereby determining its prospective value (Ariely 2009, pp. 2–3).

This is a concept that will be better clarified when we turn to the *endowment effect* (Sect. 4.5), but an illustration can be offered by noting that, as has in fact been proven, we are much more willing to make an effort to save an X amount of money for something that costs twice as much (2X) than for something that costs ten times as much (10X) even if the effort we expend and the amount we set aside are the same in either case. This shows that we subjectively evaluate losses and gains in relative rather than in absolute terms (where the reference point, in this case, is the overall price), and that the *prospect theory* adequately describes many individual choices (Magrin 1995, p. 159).

On the principle of the invariance of rational choice, the ranking of preferences among alternatives should not depend on the manner in which they are presented, but this has been shown to be empirically invalid when it comes to individual choices (Kahneman and Tversky 2000a, pp. 4–5). The relativity of the prospect implies, for example, that the variation between a null possibility and an infinitesimal possibility, or between certainty and near certainty, has a greater impact on the mind than the variation of similar magnitude between degrees of possibility that fall in the middle of the probability scale (ibid., p. 9). Some changes weigh prospectively more than others. Probabilities and the relative frequencies, and so also percentages, are statistical formats of recent invention, and for this reason they do not activate natural human statistical capabilities (Krynski and Tenenbaum 2007, p. 447). These cognitive biases show up both in the way we process information (*editing*) and in our final evaluation of the options.

As we will see in regard to *nudges* (Sect. 4.7), the theory of cognitive bias has been strongly criticized. For example, in their account of prospect theory, Krynski and Tenenbaum (2007, p. 446) argue that in certain social situations that require us to form moral judgments, whenever we are provided with a clear causal model, and statistical probabilities can be applied with ease to such a model, we are able to correctly reconstruct the distribution of probabilities and not use the representativeness heuristic. The cognitive biases that can be expected to have the most significant impact on legal behavior are the *endowment effect* and the *equity seeking* bias. These biases will receive a dedicated discussion in the following sections.

I think it opportune, however, to preliminarily analyze the problem of bias in the judicial and forensic fields (Bona and Rumiati 2013; Bona 2010; Catellani 1992; Danziger et al. 2011; Dhami 1982; Frank 1949; Guthrie et al. 2001; Pogliani 1989; Rachlinski 2000a; Rachlinski and Wistrich 2017; Forza et al. 2017; Callegari 2017). Jerome Frank (1949, pp. 104–105) wondered what the stimulus is that drives a judge to justify one conclusion rather than another. He concluded that there are individual biases constantly at work (ibid., p. 106), like the bias under which we tend to see what we would like to see, an inclination that was subsequently termed *self-confirmatory bias*.

In comparison with other professional figures, lawyers, probably owing to their *forma mentis*, may be even more disadvantaged in handling certain forms of logical inference (Bona 2010, p. 66). Judges are no less fallible in this than an ordinary witness. Not even the most demanding professional training or the most stringent selection can exclude bias from individual decision-making processes. Rachlinski (1982, pp. 136–137) notes that even the most qualified and scientifically trained individuals make serious statistical errors, and that debiasing procedures are particularly invasive and do not guarantee success. Bona (2010, p. 15) points out the mechanisms that have been worked into every judicial system as a way to make up for the judge's fallibility: these include the appeals process, taxonomies of judicial mistakes, and standards of professional responsibility.

The truth undergoes significant transformations even when examining witnesses, and this has been clear since the first systematic analyses on the psychology of testimony (Musatti 1931). This does not mean that the judge's decision-making processes are not qualitatively and quantitatively better than those of people with no legal training. Jurors routinely undervalue circumstantial evidence like fingerprints or DNA, while they overvalue eyewitness identifications and confessions (Heller 2006). Judges favor falsificationist reasoning, and indeed take a greater number of evidentiary elements into account, arriving at a much more detailed reconstruction of the facts (Giusberti et al. 2010, p. 372). But this is not in itself sufficient to ensure that they are right and immune from common mistakes. When assessing testimonial evidence, for instance, they do not have sufficient awareness of perceptual processes, nor do they always properly assess the reliability of memory (Glannon 2014, p. 163).

An unreliable witness may appear confident and convince a judge by playing on her instincts, while signs of insecurity or anxiety caused by the stressful situation can be misinterpreted as evidence that the deponent is unreliable (de Cataldo Neuburger 1988, pp. 236, 252). Add to that the ethical problem involved in realistically and reliably reenacting an event with an allegedly traumatized victim, which could trigger a process of revictimization. Beta-blockers such as propoanolol have been used in some cases, almost experimentally, on victims or witnesses of traumatic events: it has been shown that, if taken within six hours of the trauma, this antagonist drug acts on noradrenalin receptors to depress the activity of the amygdala, and therefore to dampen the emotional component of traumatic memories, thereby reducing the risk of post-traumatic syndromes. But at the same time the drug also reduces the individual's ability to remember and reconstruct an event, and so the ability to establish responsibility (Oliverio 2012, pp. 72–73).

The way in which witnesses, in some jurisdiction, are addressed when cross-examined ("Isn't it true that …") implies that a certain answer is expected, and it has been argued that relying on outward gestures and body language, especially with people with whom there is no personal acquaintance, leads to assessments of reliability lower than 50% (Cohen and Knetsch 2000), just like a purely random choice. Many errors and inconsistencies were also highlighted in decisions granting bail and in setting the amount of bail (ibid.). In a study of 167 judges in the U.S. federal court system (Guthrie et al. 2001, p. 784ff.), five common cognitive biases were found (anchoring, framing, hindsight bias, representativeness, and egocentric bias). Like many other professionals, judges make decisions quickly and in uncertain conditions, and they tend to rely more than others on heuristic shortcuts.

Judicial errors are not necessarily attributable to incompetence, negligence, or malice. They are more likely to be owed to the same mnemonic and hedonistic phenomena which afflict us all, and which have so far been investigated only in the behavioral analysis of law. The remedies so far suggested against judicial bias—such as limiting the use of lay judges or making the judiciary demographically more representative of the broader population—cannot ensure significant changes for the better, while improvements could be achieved by providing cognitive feedback on errors and inconsistent choices (Dhami 1982, pp. 69–71). Even the caps that in some jurisdictions are placed on the amount of money damages a judge can award probably reflect the awareness that it is in some way necessary to temper the decision-maker's certainties (Guthrie et al. 2001, p. 829). Yet stakeholders and lawyers, even when they are made aware of these risks, are still wary of "rational" measures and rules that automatically apply to reduce the judge's discretion.

4.5 The Endowment Effect and Equity Seeking

It was previously noted that the endowment effect is probably one of the most fundamental and influential cognitive biases in risk/opportunity assessment. The endowment effect is a concrete application of prospect theory to any decision-making that involves the risk of a loss or the potential to make a profit. The expression *endowment effect* was coined by Thaler (1980), but this bias has been an object of extensive experimental testing. The concept seemed to have already been well understood by Hume (1739, p. 251): "Men generally fix their affections more on what they are possess'd of, than on what they never enjoy'd: For this reason, it wou'd be greater cruelty to dispossess a man of anything, than not to give it him. But who will assert, that this is the only foundation of justice?" And Jhering (1913, p. 49) also had a similar insight: "No fine phrase, nor sublime speech, nor outpouring of feelings in words and tears has such convincing force as the dollar which issues from the pocket; a man's cashbook occasionally tells more concerning his true character than his diaries."

The *endowment effect* can be defined as the perception that what we possess is intrinsically more valuable than other equivalent objects that are not available to us. From this it follows that losses have a greater psychological effect than gains of equal magnitude (*loss aversion*). While most people prefer the certainty of earning 100 euros to a 50% chance of earning 200 euros, when the choice is reframed from a loss perspective, only a minority prefers the certainty of losing 100 euros to a 50% chance of losing 200 euros. From an economic point of view, the trade-off is exactly the same, but the cognitive bias of loss aversion would lead us to behave differently in the two scenarios (Guthrie et al. 2001, p. 794).

It follows from these findings that classical economic theory, which tends to forecast that transactions will take place when the sale price and the purchase price meet (once the effects of wealth and transaction costs are discounted), is not always capable of predicting when transactions in fact take place. If in ordinary commercial exchanges this effect can be modest (Kahneman and Tversky 2000a, p. 14), in an exchange of concessions—as in negotiations aimed at settling legal disputes, where the stakes are much higher—the parties involved will find it considerably more difficult to reach an agreement than would be rational to expect (Arrow et al. 1995; Mnookin and Ross 1995). The parties experience an irrational obstacle in exchanging concessions and settling their respective claims (Cohen and Knetsch 2000), and legal rules may reflect this subjective perception (Cohen and Knetsch 2000). Judges, too, in deciding who can rightfully claim ownership to something, tend instinctively to favor the possessor, even when possession may be substantially arbitrary (Guthrie et al. 2001, p. 799).

This should lead to further practical consequences: the most effective concessions to a counterparty are not those that raise the prospect of a further gain but those that reduce its losses. Depriving someone of something over which they had a simple expectation is perceived as less serious than taking away what they already possessed (Dawes and Orbell 1995, pp. 57–58). On the other hand, a business is in a way justified if, by raising prices or reducing wages, it tries to *maintain* the profits it was already making, but not if it takes the same measures to *increase* those profits (Kahneman and Tversky 2000b, p. 485).

The endowment effect is a psychological effect caused by an apparently irrational attribution of value. The classic example is an experiment in which two persons are asked to assign a value to a good of trivial value that cannot immediately be monetized, and where one person is in possession of the object (in the example a mug), while the other is not: when we are in possession of the mug, its intrinsic value—the price for which we would be willing to be deprived of it—is almost three times higher than what we would offer to buy it if we were not in possession of it (Jones and Brosnan 2008, pp. 1941–1943). The same effect can be observed in a softened form, or not at all, with goods whose "salience" is less intuitive (for example, tokens that can be used in exchange for the mug).

Initial distributions of goods tend to be "sticky," that is, they tend to remain stable even when this is not justified by economic efficiency. The same has been observed in situations involving social status: social roles or imbalances tend to be accepted as they exist. This has been noted with the theory of the belief in a just world (Lerner 1980), and the phenomenon can also be explained as an attempt to reduce cognitive dissonance (Festinger 1957). The two paradigms clearly have a common basis.

The endowment effect often manifests itself in unpredictable ways, and not always clearly: it only applies to certain goods, in certain contexts, and with some individuals more than others (Jones and Brosnan 2008, p. 1948). In some situations its incidence is greatly reduced. Studies that have attempted to quantify the proportion of affected individuals range in their estimates between 53 and 95% (Jones and Brosnan 2008, p. 1977). The effect does not seem to apply to those who act on behalf of others or those in a situation of delayed gratification (ibid., p. 1185): when the choice is between a lesser immediate benefit and a greater delayed benefit that is significantly postponed, the endowment effect becomes negligible: if given a choice between a cookie now and two cookies in an hour, we usually choose the first option, but if the choice is between a cookie in twelve hours and two cookies in thirteen, we are more likely to wait and choose the second option.

Attempts to explain the endowment effect have focused on different elements, such as the mental transaction costs required by any change in the status quo, our personal wealth at the starting point, or the strategic differences between the assignor and the assignee. The prevailing, and all-things-considered more convincing, explanation is still the one provided by prospect theory (how a choice is perceived) and by the framing theory (how the choice is presented). The effect is also a consequence of the loss aversion bias: our loss aversion is asymmetrically stronger than our attraction to gain. As to why we are so instinctively and irremediably averse to losses, the explanation may once more be evolutionary. As noted in Sect. 3.4, we are descended from an ancestral epoch when possession was not, as the phrase goes, nine tenths of the law but more likely *ninety one-hundredths* of it, such that to give up possession of something (including a status) was tantamount to putting one's own safety at risk.

All living beings, on this account, are instinctively possessive because their forebears could not afford to give up possession, not in a natural and social environment whose dangers would otherwise have made it difficult for them to survive. The relation between property, law, and survival are aptly illustrated in this passage in which von Jhering (1913, p. 49) remarks that the "purpose of life's maintenance produced property—for without property there is no secure future for existence; the purpose of the two conjoined leads to Law—without law there is no securing life and property." In fact, living beings can maximize their chances of survival if they internalize evolutionarily stable rules on property that minimize risks (Stake 2006). As previously argued, this hypothesis is not invalidated by the cultural differences between legal systems, since the coordination that every legal system is intended to make possible can be achieved in forms that may well be arbitrary. Thus, there are rules that teach us to be respectful of those who demonstrate stable possession of something.

As concerns the *enactment of property*, Collins (1987, p. 205) notes that this is an implicit aspect of social "micro-behavior," too often ignored but no less important for that. As Collins explains it, property is enacted by appropriating a physical space, but I would assume that this also includes the appropriation of physical objects. The protection of property as an assertion of domination and power does not necessarily outstrip the protection of property as a guarantee of safety from existential insecurity.

It was previously remarked that loss aversion makes it more difficult for the parties involved in a dispute to make mutual concessions (Mnookin and Ross 1995, p. 16). Everyone is intimately convinced that the counterparty is overestimating its rights. The most effective concessions (or, better yet, the less ineffective ones) are those that can be framed in terms of a reduction of losses; the least effective ones are those that improve a situation that was already framed as a gain (Kahneman and Tversky 2000b, p. 57).

The law should take this into account before concluding that the good faith with which one side supports the value of one's own right cannot be a correct criterion for reassigning that right, even in the theoretical scenario of zero transaction costs. This would explain the rarity of contractual arrangements which have great potential and are economically rational, such as reverse mortgage, but which we instinctively perceive as disadvantageous or risky. Public policies that promote modern cooperation and sociality should consider the biological constraint of our relationship to property. In this respect, Bentham was probably wrong in his assessment that rules of property are entirely artificial. Our possessiveness may be understood as a case of *time-shifted rationality* (discussed in Sect. 3.4; see Jones et al. 2001), for it now takes the form of long-term maladaptive and self-serving or self-damaging behavior. But it ought to be recognized that it cannot be eradicated without a clear awareness of how "human" it is.

Another crucial consequence of *loss aversion* is the so-called *status quo* bias (Kahneman and Tversky 2000b, p. 481; Marchetti 2008, pp. 87–88). This is the instinctive tendency to preserve an existing state of affairs, when it is one of the options on the table. Unless there are compelling reasons to change, we rarely prefer to alter the status quo. The effects of this bias is splendidly highlighted in an empirical study in the criminal field (Danziger et al. 2011). The sample of the research consisted of more than a thousand parole-board decisions in response to requests for probation. The decisions were taken over the span of several days of hearings exclusively dedicated to examining the cases. The study looked at the variables depending on which requests for probation and conditional release were accepted or denied. It emerged from the data that the most important element in determining these decisions was the *time that had elapsed since the judge's last coffee break*. In fact, at the beginning of each hearing, the rejection rate was around 65%. As time passed, the rejection rate would rapidly increase, approaching 100%: as the judges' blood sugar levels drop, the judges experience cognitive fatigue. This hypothesis appears to be borne out by the fact that when the judges took breaks (and presumably had snacks), the rejection rate would return to "normal." Meanwhile, applicants who had been examined just before a break had a much lower chance of being granted parole, and so found themselves at a disadvantage for entirely arbitrary reasons. The larger hypothesis, as the researchers saw it, is that when judges are fatigued, they tend to preserve the status quo for fear of making mistakes, which induces them to refuse parole even for prisoners who probably would have deserved it according to the standards of judgment applied in more lucid mental conditions.

The idea that judicial decisions can be explained by looking at the time the judge last snacked, as certain legal realists tend to do, has mockingly been termed the "breakfast theory of jurisprudence" (King 1937) and is criticized by Barberis (1998). In this regard, Castignone (1995, p. 287) notes that even Beccaria, in his *Essay on Crimes and Punishments*, claims that among the factors that account for the "spirit of the laws" is the judge's "good or bad digestion" (Beccaria 1764, p. 23).

In social relationships, the tendency to preserve the status quo has been described in the theory of "system justification" (Blasi and Jost 2006), which posits a tendency to rationalize this bias early on in the process. The tendency is exemplified by the behavior of disadvantaged groups, who often prefer to associate with people from other groups that are less disadvantaged than their own—this in partial contradiction of the universal inclination toward in-group thinking and conformity, that is, the tendency to prefer and favor people who are like us (Blasi and Jost 2006, p. 1121). The authors argue that in cultures where egalitarian ideals are largely contradicted by deep social inequalities, there emerge complementary stereotypes ("poor but happy," "rich but unhappy") whose main purpose is to reduce the cognitive dissonance the social situation entails for individuals (ibid., p. 1123).

System justification carries considerable implications for the law, particularly with a promotive and responsive conception of law. In certain situations, victims of discrimination or abuse may come to think it somehow justified for their rights to have been violated. Of course, this does not mean that all rationalizations of the status quo are thereby unfounded: it simply means that we have an automatic and unconscious drive to rationalize the status quo. Indeed, this makes it easier to cope with it, thereby relieving us of the anxiety and uncertainty it entails. So it should not come as a surprise that we are predisposed to resist changes in the status quo when these changes appears imminent and inevitable (Blasi and Jost 2006, p. 1143).

A further bias which indirectly reinforces our tendency to advance our own self-interest, and which is linked to the endowment effect, is the fallacy of the *zero-sum assumption*. In negotiation literature, this is often referred to as the "fixed pie" assumption (Fisher and Ury 1981). On this assumption, negotiators engage their counterparties by attributing to them interests that are symmetrically opposite to their own on the issues that are being negotiated. Thus negotiation is often mistakenly regarded as a zero-sum game, where the protection of one party's interests necessarily means frustrating those of the other. As a result, no one looks for options that may be mutually beneficial (Pietroni and Rumiati 2008, p. 194).

Specular to the endowment effect, and in some respects a counterweight to it, is the aversion to accumulation of property aimed at establishing status domination.

The widespread disapproval of this practice is not just a cultural taboo. The human species has always had an egalitarian drive, especially in the hunter-gatherer phase (Ridley 1996, p. 243). This disapproval is in a way another form of bias, and in the short term it can run counter to the aim of rationally maximizing one's wellbeing. It has been termed the *justice-seeking* or *equity-seeking* bias, and its implications are made clearer by the so-called *ultimatum game* (Nowak et al. 2000). As explained in Sect. 2.3, since the responder in the ultimatum game is in the economically weaker position of not being in possession, he holds the negotiating position of being able to respond with either a *yes*—in which case the two players will share the prize according to the division offered by the proposer—or a *no*, in which case neither player gets anything.

When the experiment is carried out between players who do not know each other, and among whom there will be no future interaction, it gives unlimited punishing power to the economically weaker player, incentivizing the other to make a more reasonable proposal. Studies on the ultimatum game have been conducted in different countries and using different forms of incentive, monetary or otherwise, and in the majority of the populations studied they consistently show that proposers will offer between 40 and 50% of the amount in their possession (Jones and Goldsmith 2005, p. 439), while about half of the economically disadvantaged players (the responders) will tend to reject any offer below 30%. This 30–40% range seems to be the universal threshold of what is perceived as a fair division. Below this threshold, the economically disadvantaged player's incentive to punish the advantaged one for a stingy offer outstrips the primary instinct to maximize utility—an instinct that should prompt the disadvantaged player to accept any offer above zero, since there is no gain to be had by punishing the proposer for a mean offer (Nowak et al. 2000, p. 1774).

When complicating elements are introduced, such as the possibility of knowing how the proposer behaved in the past, the results often turn out to be different. Even a familiarity between the players would change the outcome: in the variant of the ultimatum game called *dictator game*, the disadvantaged player is not even in the negotiating position of being able to decline the offer, but if the advantaged player receives information about the disadvantaged one (say, a name or a picture), he or she will invariably make a more generous offer (Charness and Gneezy 2008). The results of the ultimatum game have been found to remain appreciably constant across different cultural contexts, and in some cases it was surprisingly found that players in industrially developed societies, with an integrated market and institutional incentives for cooperation, were more inclined toward generosity (Caterina 2008b, p. 211).

As we saw in Sect. 2.4, when we looked at studies of procedural justice, the sense of injustice provides sufficient motivation to do acts or take initiatives that would otherwise not make sense. A clear example is that of seeking justice by bringing a lawsuit. On a purely economic-rational basis, we should think that an individual will bring a dispute to court only if the amount of damages that can be recovered, discounted by the chances of success, will be higher than legal fees. On the other hand, it is clear from experience that the desire to defeat the counterparty or inflict a punishment on him (or her) leads us to make absurd choices, driven by what has been characterized as a "taste for spite" (Jones et al. 2001, p. 1154).

When someone falls victim to an accident caused by negligence or malice, and he or his relatives claim that they are not taking legal action "for the money" but so that justice is done, or so that what happened to them should not happen to others, they are probably expressing a true intention. Current ethological theory classifies the feeling of justice among the higher types of emotions that have evolved in mammals. This emotion affords an adaptive advantage because it would satisfy a gratifying taste for punishing those who behave selfishly, or even for punishing those who are not determined enough in punishing the selfish (Aunger and Curtis 2015, p. 59). According to a naturalistic classification of the development of primary feelings, among the basic emotions developed by reptiles we primarily find the drive to accumulate resources and improve the habitat. On the other hand, mammals would have developed filial care, a predisposition toward affiliation, the pursuit of status, attraction, love, and justice (ibid., p. 60). The feeling of justice develops exclusively in a highly social group, such as human societies, where the group is very large and its members are not, for the most part, directly related to one another (Aunger and Curtis 2015, p. 55).

The egalitarian sense of justice forms the basis of a neural mechanism that can be located in the ventromedial prefrontal cortex and in the insular cortex, two regions of the brain that have proved crucial in the study of social preferences (Dawes et al. 2012, p. 6479ff.). With the use of fMRIs, Dawes and colleagues show that some parts of the brain are activated when an exchange or an egalitarian situation takes place. A continuum could be envisioned at one end of which is the propensity for appropriation and at the other end the propensity to enact leveling mechanisms (such as food sharing or monogamy). It is not certain, however, how egalitarianism connects with altruism, if not as a basic condition for evolving and maintaining sufficiently high levels of cooperation (Dawes et al. 2012, pp. 6480–6481). It is very difficult to satisfy the sense of fairness at the conclusion of an agreement. Most of the time, the parties involved have the feeling of having left something on the table, but when fairness is perceived, it leads to stable results (Mnookin and Ross 1995, p. 13).

4.6 Anchoring, Confirmation, Hindsight, and Other Biases

The impact that other cognitive biases have on social and legal relationships is widely dealt with in the literature from many points of view (Arrow et al. 1995; Jolls and Sunstein 2006; Posner 2004; Rachlinski 2000a). The *anchoring heuristic* translates directly into a homologous bias under which values tend to be largely influenced by the first anchor point (Kahneman and Tversky 1974, 2000b; Malouff and Schutte 1989; Rachlinski 2000a; Feldman et al. 2016). Negotiation positions, for example, are affected by a preexisting situation. The studies that confirm this bias are rather concordant. When required to make an estimate of the proportion of the African countries that are members of the United Nations, the participants in the study indicated a percentage that ended up being significantly influenced by another random number obtained by running a fictitious wheel of fortune. The wheel was actually designed to always return the same two numbers: when it fell on the lower number, the participants on average gave an estimate of 25%; when it fell on the higher value, their estimate was on average 45% (Guthrie et al. 2001, p. 788). So, even if those two numbers were clearly and patently irrelevant for the purposes of this estimate, they ended up significantly influencing the final result, suggesting that the estimate was in fact formed through an unconscious adjustment that took that random number as its starting value.

In the theory of negotiation (Fisher and Ury 1981; Patton 2005), the significance of the anchoring value lies in the opportunity it affords to provide objective criteria by which to justify one's requests and thus avert a rejection that would cause the negotiations to break down. The objective criterion may consist of a market value, a judicial precedent, a scientific or professional standard, cost, or any other efficiency standard, but it could also be a moral rule, a nondiscrimination principle, tradition, or reciprocity. This is referred to as the principle of arbitrary coherence: an evaluation or an initial price will influence subsequent ones without any rational basis (Ariely 2009, p. 28). The empirical studies conducted by Englich and Mussweiler (2001) show that the anchor bias exerts a strong influence even in judicial decision-making. In a series of hypothetical cases presented to criminal judges, in which the only variable was the punishment sought by the prosecution, the judges' rulings appeared to cling to those requests: this happened independently of how the judges perceived themselves to be influenced by the request, and also independently of their experience. Beyond the possible distortions that could derive from such an experimental setting, this last finding is particularly significant, for it suggests that the heuristics of decision-making play as important a role as professional habits, if not a more important one.

A universal bias that counterbalances conformism and adaptation is the tendency to *overestimate one's own judgment*. The apparent contradictions of some of these heuristics can actually be explained on the basis of the social contexts in which the decision-making takes place. In some situations, the weight of the social hierarchy is so clear and overwhelming that it is convenient to maintain a low opinion of oneself and conform to the majority. In other situations, it is individually convenient to exercise our ability to challenge mainstream thinking and impose our own way of seeing things (Wright 1995, p. 268).

Our tendency to overestimate our own judgment is described with variegated terminologies: optimistic overconfidence, egocentric bias, self-serving bias, false-consensus effect, better-than-average effect (Guthrie et al. 2001; Pietroni and Rumiati 2008). We believe that others value us more than they really do, because a sincere optimism is beneficial to us (Kahneman and Tversky 2000b, pp. 473, 476). We cling to those theories, beliefs, and expectations that align with our own, while we easily forget those that deviate from ours and introduce dissonance (Mnookin and Ross 1995, pp. 13, 17). Individuals with mild-to-moderate forms of depression generally have a more realistic assessment of themselves.

Our opinion of others depends on their opinion of us, because, while we may not realize it, it seems more coherent to appreciate those who appreciate us and to disdain those who criticize us (Blasi 2002, p. 1250). When the parties to a controversy consider the possible outcomes of a judgment, they systematically tend to estimate those outcomes as tilting in their own favor. But as soon as the judge's possible sources of doubt are pointed out to them, they revise these expectations down (Babcock et al. 1997). The effect is paradoxically accentuated even in group decision-making, where the presence of peers or experts ought to keep excessive self-esteem in check (Mnookin and Ross 1995, pp. 17–18). Self-esteem operates as a self-deception mechanism which in some situations provides the necessary stimulus to build confidence and overcome fear of the new or of failing to measure up, but which in several other cases leads to disastrous decisions, as it induces us to ignore warning signs and critical voices.

In a study of the financial frauds that were among the causes of the corporate scandals of the early 2000s, Langevoort takes the Enron case as an example of the psychological mechanism in which ego, ambition, narcissism, and self-deception, accompanied by ad hoc rationalization, induced the members of that organization to convince themselves that the rules they were violating were "artificial, arbitrary, archaic and largely unenforced" (Langevoorrt 2015, pp. 287–288). The larger picture is complicated because it was apparent that these behaviors (made known to the public) were not just the doing of sociopathic egotists solely in pursuit of power and money. Rather, they were complemented with a component of hyper-optimistic and self-deceptive behavior by managers who had become prisoners of their initial success, and were no longer able to regain control of the situation, for they could not afford the loss of professional prestige that would follow any failure to regain such control (ibid., p. 1161). In another context, Godsey (2017) illustrated how public prosecutors, owing to tunnel vision and overconfidence, tended to convince themselves of the defendants' guilt, even when evidence of their innocence had clearly emerged.

Excessive self-esteem and self-deception is probably the root of many deviant behaviors. As Langevoorrt (2015) notes, the distinction between overconfidence and arrogance is very subtle. This is even more apparent in organizations whose leaders are selected precisely because of their optimistic attitude and charisma. The financial scandals—and the public reaction to them, with the public's request for harsher penalties and stricter compliance—also provide an opportunity to consider another kind of dynamic, namely, the legislative overreaction to isolated but overwhelmingly publicized pathological events (Marchetti 2008, p. 85).

This connects to the question of the tendency of modern regulatory systems to fall victim to what Crozier (2010) calls "stable vicious circles," as evidenced by the hyper-reformism that sees public sector bodies undergoing increasingly frequent overhauls. The legislative reaction to a public outcry inevitably results in the enactment of stringent and bureaucratically burdensome disclosure requirements, which only sets the stage for a further cycle of noncompliance. A different solution that would stand on a more solid cognitive ground would consist in introducing procedural rules designed to slow down the adoption of strategic decisions whose fallout is consequential and wide-ranging (Jolls and Sunstein 2004, p. 4): this would counterbalance the tendency of management to overestimate its own capabilities.

Decision-making theory and cognitive science have also identified "defensive" and "protective" biases for which there is ample empirical evidence, showing that they affect decision-making processes with varying frequency and intensity. One of these is *reactive devaluation*, which in the first place explains why our assessment of an agreement, and any type of transaction or concession, may change even considerably depending on whether the person proposing the agreement is perceived as an opponent (Mnookin and Ross 1995; Pietroni and Rumiati 2008; Ross 1995). The very fact that a proposal is made by the counterparty seems to diminish its intrinsic value. Thus empirical research carried out during the Cold War shows that a hypothetical proposal for nuclear disarmament would be evaluated very differently depending on where it came from, even if its content remained unchanged. In particular, the same disarmament proposal received an approval rating of 90% when attributed to own's leaders, 80% when attributed to a neutral party, and 44% when attributed to the enemy (Ross 1995, p. 29).

Ross offers some strategic advice by which to overcome the impasse created by reactive devaluation. What works best is a willingness to take the first step in making small concessions, pointing out how they link directly to the values and needs expressed by the counterparty. Although these concessions come at a strategic cost, in the economy of a dispute or negotiation they can bear unexpected fruits in the medium to long term (ibid., pp. 39–41). This bias leads to two other paradoxical consequences, quite familiar to any negotiation expert: first, proposals made through a neutral party or mediator are more likely to succeed because this makes them less unsavory in the eyes of those to whom they are addressed; and, second, there is a tendency to devalue options that are already available, meaning those that can be taken without having to bargain for them (Mnookin and Ross 1995, p. 15).

Another bias that specifically comes into play in negotiation and litigation is *regret aversion*. On this theory, negotiating decisions are informed by an unconscious need not to subsequently regret having made them (Bellelli and Di Schiena 2012). The feeling of regret come in when we discover that we could have achieved more by deciding otherwise. In judicial litigation, the parties' unconscious desire to avoid this remorse would seem to push them to prefer an out-of-court settlement (Guthrie 1999, p. 43). However, this contradicts predictions that are abstractly formulated on the basis of other theories. In fact, supposing that the endowment effect is predominant, we could observe that a compromise would be strongly preferred only by those who frame the judicial decision as a possible cause of losses relative to the status quo: this would generally be the position of the defendant, who usually enjoys a strategic advantage in the contest to win a judicial claim.

Another defensive bias is the *sunk cost fallacy*. This is the predisposition to reason that since we have devoted considerable time and resources to an enterprise, we should persevere in this effort even in the face of failure, just for the sake of not allowing that past investment to go to waste (Bona and Rumiati 2013; Pietroni and Rumiati 2008). In negotiation dynamics, this bias is linked to the need to reduce cognitive dissonance (Navarro and Fantino 2009), and it involves an escalation of emotional investment (Bona and Rumiati 2013, p. 217ff.). Those who are involved in a prolonged conflict will try to minimize the inconvenience of the situation they find themselves in. A decision to stay in the fight means that we need to rationalize failed attempts to come to an agreement, and is normally accompanied by emotional hyper-involvement.

An open commitment not to compromise, and the degree to which the additional costs incurred may weigh on the terms of what could now be an acceptable settlement, cause further dissonance and frustration, which can be countered only by investing even more energy in the fight (Mnookin and Ross 1995, pp. 18–19). This type of fallacy helps to explain apparently self-defeating behavior in litigation, as well as the common paradox where certain controversies end up costing more than the object of the dispute (in behavioral economics, this corresponds to the irrational behavior of *overbidding*).

Stepping outside the realm of the parties' negotiating behavior, we have so-called *hindsight bias*: it figures among the biases that influence evaluation and decision-making mechanisms, for it affects the way we formulate judgments of responsibility (Kamin and Rachlinski 1995; Rachlinski 1998; Arkes and Schipani 1994). Hindsight bias is a "retrospective distortion of judgment" (Bona and Rumiati 2013, p. 207ff.), which means that in assessing individual responsibility for past actions, we are inclined to use a higher and more stringent standard of diligence than what we would use in other situations.

Hindsight bias is akin to the phenomenon known as *outcome bias* (Arkes and Schipani 1994, pp. 592–593). As the name suggests, this bias inclines us to look to the outcome of a decision as a measure of its quality (Rachlinski 1998, p. 581). Each person, and therefore also judges, overestimate the predictability of past events (Kamin and Rachlinski 1995), because they tend to unconsciously and automatically incorporate known events into their stock of knowledge, accordingly drawing inferences about their predictability (Rachlinski 2000a, pp. 67–68).

To illustrate, suppose that in the early hours of a misty morning a driver skidded onto the oncoming lane while cruising at 80 km/h on an interstate highway where the speed limit was 90 km/h. If we then learn from an expert report that, given that cruising speed and those conditions of humidity and low visibility, there was a 10% likelihood of that happening, we will unconsciously form the belief that the driver should reasonably have seen that coming, even if driving *under* the speed limit; and as a result, if we are put in a position to judge whether the driver is to be held liable, we will make that assessment using particularly strict criteria that transcend the limits to which our judgment ought to have been anchored. The error consists in thinking that an event was predictable simply because we know how the story ended, which distorts our perception of what can reasonably be predicted. Given how the decision-making process for establishing negligence is structured, this type of fallacy turns out to be recurrent (Rachlinski 1998, p. 571), and as a result, standards of strict liability sneak in where the law only provides for fault-based liability.

This hindsight bias may be explained as either motivational or cognitive. Among motivational explanations is the hypothesis that we want to see the world as a safer and more stable place than it is, so we delude ourselves into thinking that dangerous events are easily predicted. On the cognitive theory, the hypothesis is that our retrospective knowledge of an event would modify our perception of the world without our realizing it, invisibly convincing us that most events are actually predictable (Rachlinski 1998, p. 582).

There is an interesting application of the concept that illustrates, among other things, how the law is somewhat aware of the problem. This can be appreciated in the *business judgment rule* (Arkes and Schipani 1994; Cesiano 2013), a broad principle of corporate law now recognized in various jurisdictions, albeit with different contents and gradations. It answers the same concern to screen out ex post evaluations when determining the lawfulness of business decisions that a company's managers or executives "rationally" take in its best interest, only to later discover that they led to injury. The business judgment rule addresses this concern by recognizing that business decisions are highly discretionary and risky, and it accordingly requires that a more stringent negligence standard be met when seeking damages from corporate managers and executives, effectively removing the assessment of responsibility from the influence of hindsight bias. This rule therefore suggests that hindsight bias is in some ways an expected reaction, one that in a nonsystematic fashion has already been contrasted with the rules on the burden of proof and on due diligence (Rachlinski 1998, p. 625). For instance, under Rule 407 of the U.S. Federal Rules of Evidence, any remedial measures a defendant may have taken after doing harm, so as to make it less likely for the same episode to repeat itself, cannot be introduced as evidence to prove that the defendant was negligent in causing the harm (Guthrie et al. 2001, p. 828).

In other cases, an awareness of a bias of this kind against defendants has led to more relaxed standards of diligence, in such a way as to raise the burden of proof for the plaintiff (Rachlinski 2000a, p. 71). Thus in the jury instructions drafted by the California Bar Association we find the familiar negligence test that asks us to consider whether "a person of ordinary prudence [...] in the same situation and possessed of the same knowledge [...] would have foreseen or anticipated that someone might have been injured by [...] his action or inaction." The judge, however, is not required to instruct the jury to explicitly use this test (Rachlinski 1998, p. 604). Not even the most sophisticated debiasing procedures completely eliminate this cognitive distortion. The debiasing procedures that would be necessary to effectively compensate for the effect are probably not applicable in judicial decision-making, because they are intrusive and slow down the process, and may even introduce new opportunities for manipulating it in various ways to one's own advantage (Jolls and Sunstein 2004).

Similar problems have confronted anti-discrimination legislation, and affirmative action in particular, such as gender quotas imposed on the boardroom or racial quotas on campus. The assumption behind these kinds of rules is that the bias they are designed to counter is fundamentally conscious and deliberate basis (Blasi and Jost 2006, p. 1165), while research shows otherwise. The promotive function that law takes on in modern systems implies that judicial decisions are free from any error rooted in prejudice and that legislation actively prevents any formal or substantive discrimination based on prejudice. The issue of bias also impacts the penalties for discrimination, which itself often reflects unconscious and implicit biases. Discrimination has so far been dealt with on the assumption that it is intentional, and this may explain why antidiscrimination movements are often more interested in fighting "evil" rather than pragmatically dealing with "trivial mistakes."

Boudon (2009, pp. 55–56) observes that segregation can also result from natural individual behaviors which are not in themselves segregationist, but which in the aggregate, when compounded, and without anyone realizing it, give rise to tensions and social resistance that in turn lead to nefarious outcomes. Automatic stereotyping attributable to an oversimplified precognitive psychology is wrongly conceived as a factual error or a moral vice to be corrected (Lane et al. 2007, p. 440). Leaving aside explicit hatred, the fact is that implicit prejudice is innate (Frank 1949, p. xxii) and is cognitively efficient and beneficial (Elster 1985, pp. 158–159), because it relieves us to the problem of coping with complex problems. Therefore, we will not be able to prevent it simply by calling it out and condemning it. Add to this that we are especially prone to attribute individual behaviors to stable personality traits rather than to environmental contingencies. This bias has been termed the *fundamental attribution error* (Rachlinski 1982, p. 137).

A number of empirical studies show that implicit biases are very profound and relevant in every evaluation and decision-making process. In one experiment, individuals in two groups were tasked with solving word puzzles revolving around the idea of performance: before beginning the exercise, individuals in one group were asked to read performance-related words (such as *winning*, *competing*, *fighting*, *conquering*), while those in the control group were asked to read words bearing no relation to performance. It was found that, by this tweak alone, individuals in the first group consistently outperformed those in the second in the task at hand (Blasi and Jost 2006, p. 1128). This illustrates the extent to which we are suggestible by external stimuli, in what has been called the *primacy effect*.

Far more pervasive than could have been expected is the tendency to implicitly make negative associations on the basis of ethnicity (Lane et al. 2007, pp. 428, 432). This tendency even affects our judgment in establishing whether someone acted negligently, for even though the weight a motive carries in leading to a harmful act is independent of its moral nature, our assessment of that weight changes precisely on that basis, depending on whether we find the motive morally acceptable or reprehensible. In a study reported by Sood (2013, p. 311), for example, the respondents were asked what they believed to be the ultimate cause of a car accident involving a negligent driver: if the personal motive behind the negligent behavior was disapproved of (e.g., the driver was hiding a stash of drugs), the respondents were more likely to attribute the crash to the driver rather than to an external cause (such as a patch of motor oil on the road or foliage covering the road signs), but not so if the motive did not carry this stigma (e.g., the driver was trying to hide a gift in time to surprise her parents on the way to their anniversary celebration).

Implicit attitudes are traces of past experiences that mediate our behaviors, and it is important to point out that implicit biases do not in any strict sense determine behavior (Lane et al. 2007, p. 437). Implicit biases are, however, context-dependent and highly malleable (Greenwald and Hamilton 2006, p. 963), so much so that when we are exposed to individuals who contradict a stereotype (women who are physically strong or in leadership positions), our bias is immediately reduced, and the effect lingers for some time thereafter (Lane et al. 2007, p. 438).

Furthermore, the stereotype cannot always be offset by giving discriminated individuals greater representation or decision-making powers. In this case, we still find the instinct to justify the status quo or buy into "just world" theory, because disadvantaged people often carry the same prejudice against members of their own group. The remedy of exposing biased people to counter-stereotypical stimuli risks rebounding on those very individuals, only to reinforce their implicit bias on the next occasion (Blasi 2002, p. 1253).

The bias also survives through adaptation strategies such as enacting stereotypical subtypes that in particular cases legitimize the exception to the stereotype. For instance, immigrants who succeed in some profession or join the intelligentsia only offer confirmation that they are an exception to the general rule. The solutions to these problems can essentially be of two types: counterbalancing through either procedural rules (e.g., cooling-off periods) or substantive rules (e.g., positive actions).

4.7 Nudging[1]

Behavioral science is increasingly influencing fields that have traditionally fallen within the purview of the social sciences. In 2017, the Nobel Prize in economics was awarded to the behavioral economist Richard Thaler, speaking to the fact that we now recognize the utility and relevance of cross-fertilization among traditional disciplines such as economics, psychology, law, and cognitive science. A few years earlier, Richard Thaler, working with Cass Sunstein, a jurist, introduced the successful idea of "nudging" (Thaler and Sunstein 2009) in order to promote a new style of regulation inspired by a philosophy of "libertarian paternalism" (Thaler and Sunstein 2009).

Nudging lies outside the traditional "command and control" attitude to regulation, but it also does not entail wholesale deregulation: it seeks to guide human behavior by leveraging biases and heuristics that characterize the way individuals gather information and make decisions. Nudges intervene and guide decisions when the instinctual decision-making system (system 1) prevails over the properly rational system (system 2) (Kahneman 2011). In this way, nudging can become an alternative to other forms of traditional regulation, in such a way as to lower the cost of channeling social behaviors (Baldwin 2014, p. 831). As a term of art, the verb *nudge*

[1] This section is an adaptation of "Framing Choices to Influence Behaviors: A Debate on the Pros and Cons of Nudging," published in Diritto e questioni pubbliche 1(2018).

is used in policymaking in its commonly recognized meaning where the idea is to "gently push" or "elbow" someone into doing something, through an exhortation they can easily and intuitively understand and appreciate.

The point of nudging, in other words, is to guide the behavior of individuals without violating their freedom of choice. In order to convey this idea, Thaler and Sunstein (2009, pp. 5–6) introduce the example of a school canteen whose policy is to improve the students' eating habits, and to this end, instead of banning or eliminating junk food, the staff places the healthier food at eye-height (Thaler and Sunstein 2009). For widespread examples of nudging, we can also point to the nutrition-facts label that is mandatory for certain foods, the tobacco packaging warning messages, or opt-out social security plans (Sunstein 2017, p. 19).

When it comes to encouraging socially desirable behavior through better information, it is not advisable to tell someone that they are more virtuous than average, because their behavior will end up getting worse (Thaler and Sunstein 2009, p. 74). Thus, for example, people seem more willing to pay the taxes they owe when they are informed that a good percentage of other people regularly pay their own taxes and do not incur any fines, but not when they are informed about the way their taxes are used, or about penalties for tax violations, or about what further assistance is provided for filling their tax returns. Likewise, voter turnout generally does not increase by lamenting the fact that it is low (Thaler and Sunstein 2009, p. 72). The delimitation of nudges seems to be of particular interest in health care and social care, where the decisions that individuals make now not only affect their personal wellbeing in the future but have major implications for public spending.

To qualify as such, a *nudge* measure should not imply significant material incentives or disincentives: levying taxes, offering subsidies, or depriving persons of their freedom cannot be considered a nudge (Sunstein 2015, p. 512). Nudges work because they provide information effectively, because they make certain choices easy, or simply because they leverage individual inertia. In several cases, nudges seem to work by leveraging a cognitive bias. But choices can also be influenced by making information available when no bias is at work (so long as we take it that ignorance itself is not a bias) (Sunstein 2015, p. 513). Nudges aim to provide a decision-making architecture based on empirical evidence, and to overcome the moralistic paternalism inherent in much legislation.

The paternalistic route ignores biases that cause harm to oneself and implements harsh penalties for biased behavior that harms others. But this no longer seems an acceptable choice for governments. It is only in a relatively recent past that question of the interaction between cognition and human behavior came into its own as an object of sustained study, with the concepts of prospect and framing (Kahneman and Tversky 1974, 2000a, b), at which point policymakers also began to take a fresh look at science as a source of guidance in regulating society. Considering the cost and varying degree of effectiveness of traditional regulation based on threats and punishments, it may now seem appropriate to consider the alternatives.

When legislative committees seek expert opinions in the process of drafting a bill, they turn to economists, jurists, sociologists, life scientists, and so on. These committees are now supported by agencies or units mainly composed of behavioral scientists. Among the first and best-known examples are the Behavioral Insights Team (BIT) in the U.K. and the U.S. government's Social and Behavioral Sciences Team. It has been recently estimated that in some fifty-one countries, the behavioral sciences have an official policymaking role.[2] The trend has also taken hold in the international arena, considering that the World Bank and the United Nations would like to apply the principles of behavioral science to sustainability and development policies, for example in the effort to fight corruption. In its latest annual report, the U.K.'s Behavioral Insights Team claims that it has reduced the flow of patients sent to overcrowded hospitals, has limited unnecessary gas consumption, and has reduced the number of speeding tickets, while increasing the ratio of disadvantaged students admitted to prestigious universities and the number of workers enrolling in voluntary retirement plans.[3] Circumscribed applications of the nudge concept have been analyzed in specific fields like freedom of contract (Mathis and Burri 2016), consumer protection (Rangone 2012), organ donation (Den Hartogh 2013), financial investment (Helleringer 2016; Tereszkiewicz 2016), and dispute resolution (Watkins 2010).

By changing the choice-framing architecture with nudges, that is, by shaping the frame through which choices are presented, it should be possible to achieve results that are less socially problematic. The structuring of options and the use of framing seem more effective in situations where individuals have no feedback as to the outcome of their actions or where the effects of choices are deferred over time (Thaler and Sunstein 2009, pp. 83–84). An effective strategy in this respect is to set default options that one must be able to get rid of in a very simple way. Two examples are organ donation and enrolment in voluntary social security plans: in most jurisdictions, these are choices that require express consent, and not surprisingly, these behaviors are not as common as we want them to be.

If we could presume these virtuous choices, we would be able to significantly improve outcomes simply by making it possible to opt out with a mere statement. Other effective illustrations of nudges are the default setting on computers, mobile phones, and GPS navigators (Sunstein 2017, p. 1). These settings can be used to guide and direct behavior while ultimately giving individuals the freedom to bypass them. They are "institutional prosthetics" that can save lives, as in the case of the parallel road stripes perpendicular to the flow of traffic: they give drivers the impression that they are driving faster than is actually the case, thereby inducing them to slow down (Trout 2005, p. 427).

The rationale behind nudging, and the reason why Thaler and Sunstein chose for it the oxymoronic label "libertarian paternalism," is that they are not meant to *replace* traditional regulation but to *complement* it, in such a way as to preserve each citizen's agency and control. Liberal paternalism is an attempt to devise a "third

[2]Assessing the Global Impact of the Behavioural Sciences on Public Policy (2014): https://changingbehaviours.files.wordpress.com/2014/09/nudgedesignfinal.pdf, accessed on Oct. 27, 2017.

[3]The Behavioural Insights Team Update Report 2016–17: http://38r8om2xjhhl25mw24492dir.wpengine.netdna-cdn.com/wp-content/uploads/2017/10/BIT_Update-16-17_E_.pdf, accessed on Oct. 27, 2017.

way" (Giddens 1998) in regulation. In the midst of a polarizing public debate, nudges seek common ground in the kind of gradual reform that may appeal across a broad political spectrum (Amir and Lobel 2008, p. 2118). Contrary to nudging, "command and control" regulation and total deregulation often entail, in equal measure, a drastic overhaul of the existing framework, and so they end up with measures that are either outright coercive or fully libertarian (Sunstein 2000). The idea that legislative intervention should be gradual and noninvasive, while recognizing that people sometimes act in self-damaging ways, has been expressed in various forms. One description is "paternalistic," sometimes qualified, as in *asymmetric paternalism* (Whitman and Rizzo 2015, p. 410), and another one is "debiasing through law" (Jolls and Sunstein 2004).

As the previous discussion may have suggested, nudging seems to ultimately manipulate our will without our being able to realize it. An effort has therefore been made to understand how it might be received across different cultures, and it seems to have a strong reception in most countries, with similar patterns found in the United States, Canada, Australia, the United Kingdom, Germany, France, and Italy (Sunstein 2017, p. 17). Sweden has been observed to be more receptive to it than the United States, and it has been speculated that this may well have to do with Sweden's national culture of wide support for government intervention (Hagman et al. 2015, p. 449). Slightly lower levels of acceptance have been recorded in Denmark, Hungary, and Japan, while acceptance in Italy and the United Kingdom is high compared to other European countries (with the exception of "sweets free" cashier zones in supermarkets in Italy) (Sunstein 2017, p. 34). Nowhere in the world does nudging find greater acceptance than in China and Korea (ibid., p. 18). Acceptance of nudging does not seem to vary with demographic, but gender does seem to be a factor, women generally being more in favor of it than men (ibid., p. 36).

A significant difference affecting nudge evaluation was linked to the nudge's educative potential. Noneducative nudges are constructed by reframing choices without making people better informed: they rely on the instinctive decision-making system, as they play on people's fears or hopes (Sunstein 2017, p. 7). Information-giving or educative nudges tend to win greater acceptance, but there is also remarkable variance from case to case, depending on how effective the nudge is (ibid., p. 9). In short, whether a nudge will find favor depends not so much on its abstract proposition as on how effective it proves to be in practice.

As suggested, the idea of exploiting biases and heuristics to achieve socially optimal results is certainly subject to objection. Two mains lines of criticism have developed, each with its own range of views: on the one hand we have the political and ideological criticism that object to the underlying philosophy of libertarian paternalism; on the other hand, we have the psychological-cognitive criticism that challenges the validity of the cognitive bias theory or its practical use as a criterion on which basis to tweak regulation.

In the first camp we find critics who argue that there is little in libertarian paternalism that is actually libertarian and much that is manipulative. Thus Baldwin (2014) has highlighted some forms of nudging that, instead of relying on a requirement to keep the public informed about its object or on the inertia that causes us to cling to default rules, conceals a paternalistic approach that leverages our emotions. These are characterized by Baldwin as “third degree” nudges, the prime example being the fear-inspiring labelling on cigarette packaging, considered to be explicitly manipulative: for this reason, and because there is no opt-out, this kind of nudging can be said to break the trust between regulators and the public (Baldwin 2014, p. 836). When these measures are taken directly by administrative agencies, without the public debate that accompanies the legislative process, nudging could even be a form of “disrespectful social control” (Hausman and Welch 2010, p. 134). Similarly, Mitchell (2015, pp. 1247–1248) argues that nudging are ultimately a form of paternalism in disguise, particularly when designed for wealth redistribution, for the underlying assumption is that the wealth is redistributed from *rational* individuals to *irrational* ones.

As several critics have also pointed out, nudging assumes that decision-making bias only affects the people being nudged, not the people who are *doing* the nudging. Nudge theory, in other words, seems to imply the existence of a benevolent and rational public official in charge of nudging, but where is this rational being to be found? (Niskanen 1971). There is no reason why government action should be immune from the same cognitive errors as individual action (Lodge and Wegrich 2016, p. 253): it, too, is subject to the constraints of “bounded rationality” (Simon 1957). The anarchy and messiness of government decision-making have been widely underscored in the literature, where government is compared to a “garbage bin,” from which good plans and measures can get discarded and then recovered by mere accident or for the wrong reasons (Cohen et al. 1972). Libertarian paternalism is argued to be in some ways a violation of the libertarian and individualistic principle of *value neutrality* (Amir and Lobel 2008, p. 2120; Steffen 2016, p. 84). In the same vein, nudging has been described as “welfarist” (Guala and Mittone 2015, p. 386). Likewise, and also in seeming contradiction to the principles of libertarian paternalism itself, some nudging has been described as pro-*social*, one example being the nudging designed to promote recycling, another the opt-out rules that by default allocate part of an individual’s income to charity. These measures are “epistemologically different” from pro-*self* nudges (Hagman et al. 2015, p. 441), which are only designed to help us make better choices in our own self-interest. This would make pro-self nudging less intrusive than pro-social nudging, and therefore more acceptable.

The second line of criticism is the psychological, with several authors challenging the theory of systematic cognitive bias (Arrow et al. 1995; Kahneman and Tversky 1974, 2000a, b). Here Gigerenzer (2015) raises a number of objections that point out the inconsistencies in the very concept of cognitive bias, and he subscribes to the notion that when people are well informed and have been educated about a subject, they will make fewer irrational choices.

But the main problem with libertarian paternalism, according to Gigerenzer, is that it implicitly places all blame on the individual, while ignoring that institutions may have a vested interest in diverting individual behavior in one way or another. This objection takes issue not with nudging itself but with its use as a convenient shortcut that is taken in place of the longer and more resource-intensive route of educating those we want to nudge (Gigerenzer 2015, p. 363). Although nudges are to be credited for making bureaucrats aware of the psychological factors at work in the decision-making process, they can become an excuse for not adequately protecting the rights of people who need such protection, like consumers, and coercively enforcing such rights when needed. Gigerenzer also notes that the definition of nudging has become confusing, even at the hands of its own creators, who initially *excluded* from the concept any measure that works by providing material or monetary incentives, but then stretched the same concept to *include* such measures within its scope (while qualifying this extension of scope by requiring that the incentive needs to be small).

From the point of view of behavioral psychology, Gigerenzer argues that the kind of cognitive psychology and behavioral economics at work in the libertarian paternalism of nudging (Mathis and Tor 2016) proceeds from a caricaturized view of human nature and individual intelligence as irremediably error-prone. On the contrary, he maintains, it is premature to be pessimist about the difficulty of debiasing: with adequate visual and numerical support, even children can be educated about risk and uncertainty (Gigerenzer 2015, p. 364). According to Rachlinski (2002, p. 1212) we need not even resort to nudging as a remedy against cognitive errors: "simple experience" will suffice. In fact, the evidence that we are systematically irrational is far from conclusive, for it is based on tests which use particularly restrictive rules of logic, and which therefore do not accurately capture the "ecological nature" of rationality.

These tests are selectively reported through a "confirmation bias" that has virtually eliminated any contrary evidence. The cognitive biases the libertarian paternalists refer to are therefore unsystematic, rather than systematic (Gigerenzer 2015, p. 371). And in any case, the "errors" rooted in natural heuristics (availability, overconfidence, hindsight, self-serving reasoning, framing, status quo, anchoring, and adjustment) would not necessarily be worse than other logical or statistical models presented as rational.

Then, too, it is not just in cognitive psychology that we seem to cherry-pick the evidence for a fundamental irrationality: the problem is endemic across all the social sciences. In fact, the suspicion is that the published research finding that we are prone to systematic cognitive error actually represents only a small percentage of the total. The vast majority of research remains unpublished in the drawer (this is known as the file-drawer problem), precisely because it does not show any error, and so it does not point to an original discovery. This is a long-debated problem in science, where research that lacks visibility, because unoriginal or "uninteresting," may offer evidence that some otherwise well-established hypothesis is false, and that the null hypothesis is true. In fact the argument has been made that we should have publications specifically devoted to research that does *not* prove anything new (and some years ago a *Journal of Articles in Support of the Null Hypothesis* was launched). This kind of research is also necessary if science is to get rid of false scientific hypotheses that might otherwise survive until somebody accidentally "proves" them, perhaps using processes that cannot be replicated. By the authors' own admission, the experiments carried out to prove that our decision-making heuristics are fundamentally irrational were misleading by their very construction, for they did not reflect the real conditions in which decisions are made (Mitchell 2002, p. 1973). Indeed, the research that has been left in the drawer may prove that most people in most cases are capable of deciding in a substantially rational manner (Mitchell 2002, pp. 1966–1967).

According to Gigerenzer, the dualistic model of reasoning (system 1 vs. system 2) discovered by cognitive psychologists achieves clarity at the cost of oversimplification: even Freud's three-pronged model (based on ego, superego, and es), now criticized as out of date, looks more sophisticated than the dualistic system. We would probably fare better by investing resources to educate people about risk, as early as possible, and in any case before they begin to adopt risky behaviors (Gigerenzer 2015, p. 379). Educational measures would help to alert individuals to the existence of framing or other socio-psychological factors that influence our decisions (Lecouteux 2015, p. 407). Finally, there is empirical research based on randomized trials that seems to contradict the alleged effectiveness of some of the nudges cited by Thaler and Sunstein (Stevens-Simon et al. 1997).

Nudges, however, have always existed in some form, and the fact that they are now being debated so intensely perhaps testifies to an availability bias that may be at work in the debate on the nature and purpose of legislation. Never has such a wide-ranging assortment of rules, laws, and regulations been grouped under a single category, making it the target of criticism it might otherwise never have attracted. Gigerenzer reports an interesting example that regards voluntary screening for breast cancer. In some countries, women over the age of fifty receive a letter of invitation to show up for a voluntary mammography at a set date and time. Over a ten-year course, this screening reduces the mortality rate from five women in 1000 to four. In the letter, this difference is reported as a risk reduction of over 20%, often raised to 30% to make it more convincing (Gigerenzer 2015, p. 362). There is no doubt, therefore, that here we have an instance of manipulation based on statistical framing. Equally beyond doubt, however, is that the value of saving that single human life justifies the cost of sending 1000 letters, and also the time that healthy women devote to screening.

Also questionable is Mitchell's assertion (2015) that nudges would, unbeknownst to us, redistribute wealth from people who are more rational to ones who are less so. Even if irrational people are induced to eat healthier, this does not necessarily mean that they will eat less: the outcome may well be a rise in prices at the canteen, at the expense of the rational people who eat less. It is true, however, that rules requiring information to be made available to consumers may entail additional costs, taking up more administrative time and resources. These costs are also borne by rational choosers, but as long as they remain low, the social burden is probably acceptable. Even the costs of illness and bad eating habits end up falling in part on society. And Hagman's (2015) distinction between pro-social and pro-self nudges seems fictitious: nudges that apparently are pro-self may, in the aggregate, wind up making for greater social utility.

In response to some of these objections, Mills (2015) suggests adopting three criteria that nudging needs to meet if it is to be respectful of personal autonomy: it has to be (1) consistent with the authentic goals of a mature individual; (2) easily avoidable; and (3) adequately advertised and easily recognizable for what it is. At least three types of nudges would meet this standard: personalized default rules, choice prompts (provided that they do not become overwhelming), and the provision of "framed information" (ibid., p. 502). Also in response to the same objections, Wilkinson (2013) suggests making a distinction between manipulative and nonmanipulative nudges, but he concludes that this distinction is actually quite difficult to make, and that it is more productive to ensure an effective opt-out that is not merely formal, and therefore very simple to achieve.

Furthermore, even when nudges are not framed as rational methods of persuasion, they can still be used in two ways. First, they can call attention to a problem or provide information about a choice that we will have to make in one way or another (such as deciding whether or not to smoke), but here they work only so long as (a) the choice is presented at the beginning of the decision-making process, (b) the information can be easily recognized, and (c) it is not intended to shock or disturb but is nonetheless clear in its persuasive intent. Second, they can be used to support subjects whose autonomy is compromised, like someone who is unable to correct cognitive errors (Sawicki 2016, p. 212). These criteria would exclude subliminal messages, for example, or the very graphic images that are now standard on packets of cigarettes, but they would allow placing cigarettes out of prominence in retail, while giving visual prominence to devices intended to help us stop smoking (Sawicki 2016, p. 222).

A further criterion by which to justify the use of nudges lies in the process through which they are created: if the process is legislative, the nudge will have greater justification, because this means that it was subjected to a level of public scrutiny it would not have received under a nonlegislative procedure, whose outcome would accordingly not be easily recognizable. Such a nudge would thus be liable to criticism as a device that could undermine an individual's capacity for autonomous decision-making (Lepenies and Małecka 2015, p. 435).

Sunstein's final answer to some of these objections is that autonomy would not be endangered even with nudges that do not set out to achieve their objective by informing and educating. These nudges free up two of the scarcest resources we have today, namely, *cognitive energy* and *time* (Sunstein 2014). This overrides most critical arguments, on the basis that to compel a choice on technical issues is to engage in paternalism. In fact, the only possible alternative would consist in imposing a legal obligation, thereby transforming the nudge into a coercive norm of the conventional type (Sunstein 2017, p. 89). It may be advisable, then, to rely on a system that is similar to what happens when deciding how we want a software application to work, that is, when we are asked to deliberately choose whether we want to keep the default settings—as laid down by a superior and enlightened body whose judgment we decide to trust, because we do not feel competent enough in the matter at hand or we do not have the time needed to figure out what is best for us—or whether we want to customize these settings ourselves and in detail. This option could be termed "simplified active choosing." Libertarian paternalism, then, will be acceptable when instead of *requiring* us to make choices it *promotes* these choices (ibid., p. 106).

It is still too early to say whether nudging can be abandoned for something better or better-sounding. Policies promoting nudges recognize that regulation is multifaceted: it faces a range of different problems today, and so it needs to deploy a range of different tools. Nudging is therefore not a conceptually strict formula but a pliant tool, so much so that it has sometimes been stretched excessively, becoming a victim of its own success. In this respect, its creators have accepted the risk of not promptly intervening to narrow down the concept. The debate we looked at in outline does not solve the fundamental disagreement on the issue, whose sticking point lies in the proper understanding of individual autonomy and collective welfare, and what the boundary and interaction between them is.

Regardless of how those issues are worked out, however, it is clear that nudging can help regulation pragmatically address the need to take account of the different cognitive abilities of individuals (Rachlinski 1982, p. 126). So, too, on two points I would strongly agree with the authors who developed the concept: they note that nudging is unavoidable, because there is no way to avoid the choice architecture, and hence the framing; and they also note that there is no such thing as a neutral frame, for which reason it is necessary to always be aware of the frame within which a nudge is designed, while also making sure that the framing process is open, in such a way that the frame is clear to everyone and is open to revision. I believe this is already a possibility, at least to the extent that the rights, interests, protections, and principles we want to secure—such as health, savings, education, and an adequate standard of living—are explicitly stated and ranked in a constitutional charter. Nudging should therefore not be considered an exception to "constitutionally based decision making" (Frey and Gallus 2016, p. 18), nor is it inconsistent with democracy, because even when the constitution does not provide the blueprint needed to design a proper nudge, there is nothing about a nudge that should prevent us from looking to other authoritative sources when designing one, or from making the design process open and public.

4.8 The Way Forward for Cognitive Law

The cognitive sciences, finally, offer a systematic perspective on those psychological phenomena that shape our perception of the law and hence the attitude through which we approach the law and relate to it. Some empirical experiments consistently support the hypothesis that individuals would recall normative social information (and therefore social norms) better than nonnormative social information (O'Gorman et al. 2008, p. 76). This is a shared interest among all the disciplines that presuppose a theory of social behavior, and for which it is important to establish a relation of mutual influence among rationality, intuition, and emotion. In this sense, it is now possible to speak of a social neuroscience that deals with the ways in which the brain guides social behavior, and how the social world in turn affects individual biology (Wexler 2008, p. 85ff.). The idea is also emerging of a "social brain" as the sum of neural correlates and abilities directed at interactions, thoughts, and individual feelings (Forza 2010, pp. 220–221). This also brings back the debate on emotion (Wentworth and Yardley 1994), an element that is difficult to place in the framework of decision-making theory. At issue is the question of whether emotion is a sophisticated cognitive process or a complex adaptive response used as "a 'compensatory mechanism' for the 'phylogenetically driven' low sociality of humans as evolved apes" (Williams 2001, p. 53).

At least three fundamental functions have been hypothesized in regard to the emotions: to gain information about the opportunities of the outside world, to draw our attention to dangers and the occasions on which they may arise, and to guide us in making good strategic decisions (Mazzocco 2008, p. 161ff.). At the foundation of the "innate sense of justice" Fuselli (2014, p. 132) places certain basic affections such as the sense of ownership and the instinct of equality. It is therefore not only for the sake of novelty that we are now speaking of a "cognitive theory of law" (Ambrosino 2014), or that intuition is seen to also play a fundamental role in "legal cognition," where it is more solemnly described as consciousness (Sartor 2005, p. 134).

Even the alleged "neuromania" that the contemporary social sciences has pointedly been criticized for cannot in any way challenge the advances that have been made in the effort to understand individual cognitive effects. Under the ontological distinction that has been drawn between factual objects, data, and the phenomena involved in constructing meaning out of those objects and data, neuronal phenomena would be incommensurable with social ones (Marrone 2011, p. 55). But this distinction is as shaky as the philosophical critique that ignores the degree to which complex social behaviors are multifactorial, and equally shaky is the rather indiscriminate charge that the social sciences are engaged in a naive behaviorist reductionism (ibid., p. 52). That the factors which mentally interfere with our legal rationality are complex, and that they pile up in a steady stream, had already been fully understood by Frank when he wrote of "jungles of the mind," pondering whether new discoveries would at least reduce the "chaotic state" of the law (Frank 1949, p. 113), without having to cling to the "infantile hope of absolute legal certainty" (ibid., p. 159).

The hopes he did harbor turned out to be far too optimistic, because the social sciences lagged behind. On the specific contents of Frank's analysis, Bobbio one time famously offered some critical remarks doubting the ability of the psychoanalytic approach to explain our obsession with certainty (Bobbio 1951, p. 151). As much as there may be good reason for such skepticism, it does not warrant the criticism he laid upon the naïveté of such scientistic positivism. We cannot today ignore the fact that judges, just like the rest of us, are liable to "patterns of social functioning" that condition their decision-making activity (de Cataldo Neuburger 1988, p. 335), and one can only hope, with Frank, that these patterns are brought to light, since only in this way can the judge become "more sensitive, more nicely balanced, more subject to his own scrutiny, more capable of detailed articulation" (Frank 1949, p. 143).

This should not be taken to suggest that we ought to ignore the different critiques made of the cognitive theory of law, starting from the criticism that Hillman (2000) devotes to behavioral decision theory. According to Hillman, the psychological effects described as cognitive biases are presented in such an inconsistent and context-dependent manner that they cannot be restated as general laws about the real world, and for the time being they amount to no more than a contradictory "shopping list" of overlapping and sometimes conflicting effects that would lead to inconsistent policies. In fact Hillman (2000) argues that a theory of decision-making behavior founded on biases would not offer the law any clear prescription.

Similarly, Mitchell (2002) circumscribes and downplays the effects of cognitive biases, arguing that we should instead focus on the interpretive and methodological limits of the psychological research that legal behaviorists have artfully used to make generalizations about the real world. He warns us that we should not be using the experimental method to draw far-reaching conclusions, especially where a bias is detected whose effects fail to reach a statistically significant threshold and therefore cannot be used to refute the assumption of a generalized rationality (Mitchell 2002, p. 1908).

These criticisms undoubtedly have some basis. It was also previously noted that while some biases are *defensive*, others are *assertive*. The question could therefore be asked whether we are subject to conflicting drives. This concern has been expressed by noting that the same kinds of contradictions can also be found in proverbs and common rules of conduct (Haselton and Nettle 2006, p. 47). To some extent, however, these contradictions are inherent in human psychology, which, as noted, is closely dependent on environmental stimuli, precisely because it has developed in ductile ways.

I would suggest that these critiques can be sorted into two groups: one methodological and the other axiological. In response to criticisms of the methodological kind, we should reply that one task of the cognitive theory of law is precisely to fully understand in what contexts and under what conditions biases operate. Rachlinski and Jourden, for example, note that the intensity of the endowment effect in litigation depends on whether it is an injunction or a monetary award that the law prescribes as a remedy by which to make the aggrieved party whole (Rachlinski and Jourden 1998). In fact, the endowment effect comes into play only when the assets we stand to lose matter to us: if we are willing to relinquish them in order to assert a right, it means that we have already taken up a perspective of expected utility. A theory developed as recently as the cognitive theory of law cannot be faulted for being unable to describe the totality of observed phenomena. Otherwise, we would also be forced to reject the validity of most theories that govern the physical world.

The second line of criticism, the axiological, proceeds from the concern that cognitive science cannot provide the law with any sort of unified and broad support in the interests of society at large. In reply to this kind of objection, one can only observe that there are numerous ways in which the legal system can stand to benefit from the suggestions offered by the cognitive theory of law, if only we accept its overarching tenet, namely, that each individual reasons according to a different rationality. As we have seen, many are willing to pay the cost of a suboptimal outcome in negotiation or in litigation in order to avoid the risk of regretting a choice that has changed the status quo (Rachlinski 2000b, p. 759). This is a highly subjective individual rationality, and we cannot expect cognitive science to provide the law with unambiguous, across-the-board recommendations. It will all depend on the degree of complexity and sophistication we will accept to introduce into the legal system. If a legal order is to be responsive to individual differences, it will have to introduce corrective measures to take account of different perceptions of rationality.

No legal cognitivist is actually unaware that the practical effects of biases do not completely subvert the "rational" dynamics of our daily intercourse. This awareness is often incorporated in judicial practice, and this is why many jurists are dismissive of neuroscience (Fuselli 2014, p. 18). And indeed, as I believe the foregoing discussion illustrates, we should recognize that neurolaw and legal cognitivism are not yet ready to make a revolutionary contribution to law (a point we can make by going back to Levy 2014). Once we recognize as a mereological illusion the idea that the human mind and human experience are reducible to the brain's neural functioning (Pardo and Patterson 2013; Sartor 2005, p. 18), we can appreciate the need to avoid using cognitive science as a single, overarching, all-controlling paradigm by which to explain all legal phenomena (Goodenough and Tucker 2010, p. 66). But this is not to say that cognitive science cannot play any role in clarifying the legal phenomenon: an awareness of our cognitive mechanisms still enables us understand the reasons behind our own actions in ways we might not otherwise be able to appreciate, and the science can itself provide a broad framework within which to understand our own action in relation to the law.

References

Alces, Peter A. 2018. *The Moral Conflict of Law and Neuroscience*. Chicago, London: University of Chicago Press.

Aleo, Salvatore. 2010. Complessità e responsabilità penale. In *Neodarwinismo e scienze sociali*, ed. Roberto Vignera. Milano: FrancoAngeli.

Almodóvar, Miguel Ángel. 2015. *Intestino, secondo cervello*. Milano: Vallardi.

Amato, Salvatore. 2013. A chi appartengono i miei pensieri? Neuroscienze e diritto di proprietà. In *Il diritto nelle neuroscienze. Non "siamo" i nostri cervelli*, ed. Laura Palazzani and Roberto Zanotti. Torino: Giappichelli.

Ambrosino, Angela. 2014. A Cognitive Approach to Law and Economics: Hayek's Legacy. *Journal of Economic Issues* 48 (1): 19–48.

Amir, On, and Orly Lobel. 2008. Stumble, Predict, Nudge: How Behavioral Economics Informs Law and Policy. *Columbia Law Review* 108: 2098–2137.

Anolli, Luigi, and Fabrizia Mantovani. 2012. *Come funziona la nostra mente*. Bologna: Il Mulino.

Appelbaum, Paul S. 2007. Law & Psychiatry: The New Lie Detectors: Neuroscience, Deception, and the Courts. *Psychiatric Services* 58 (4): 461.

Ariely, Dan. 2009. *Predictably Irrational*. New York: HarperCollins.

Arkes, Hal R., and Cindy A. Schipani. 1994. Medical Malpractice v. the Business Judgement Rule: Differences in Hindsight Bias. *Oregon Law Review* 73: 592–593.
Arrigo, Bruce A. 2007. Punishment, Freedom, and the Culture of Control: The Case of Brain Imaging and the Law. *American Journal of Law & Medicine* 33: 457–482.
Arrow, Kenneth, Robert H. Mnookin, Lee Ross, Amos Tversky, and Robert Wilson, eds. 1995. *Barriers to Conflict Resolution*. New York, London: W.W. Norton & Company.
Aunger, Robert, and Valerie Curtis. 2015. *Gaining Control. How Human Behavior Evolved*. Oxford: Oxford University Press.
Babcock, Linda, George Loewenstein, and Samuel Issacharoff. 1997. Creating Convergence: Debiasing Biased Litigants. *Law and Social Inquiry* 22 (4): 913–925.
Baggio, Guido. 2015. *La mente bio-sociale. filosofia e psicologia in G.H. Mead*. Pisa: Edizioni ETS.
Baldwin, Robert. 2014. From Regulation to Behaviour Change: Giving Nudge the Third Degree. *The Modern Law Review* 77 (6): 831–857.
Barberis, Mauro. 1998. *L'evoluzione nel diritto*. Torino: Giappichelli.
Barkow, Jerome H., Lela Cosmides, and John Tooby. 1992. *The Adapted Mind. Evolutionary Psychology and the Generation of Culture*. New York, Oxford: Oxford University Press.
Basile, Filippo, and Giuseppe Vallar. 2017. Neuroscienze e diritto penale: le questioni sul tappeto. *Diritto Penale Contemporaneo* 4: 269–289.
Baskin, Joseph H., Judith G. Edersheim, and Bruce H. Price. 2007. Is a Picture Worth a Thousand Words? Neuroimaging in the Courtroom. *American Journal of Law & Medicine* 33: 239–269.
Beccaria, Cesare. 1764. *An Essay on Crimes and Punishments, with a Commentary by M. de Voltaire. [1872]*. Albany, NY: W. O. Little & Co.
Beecher-Monas, Erica, and Edgar Garcia-Rill. 2015. Overselling Images: fMRI and the Search for Truth. *Marshall Law Review* 48 (3): 651.
Bellelli, Guglielmo, and Raffaella Di Schiena. 2012. *Decisioni ed emozioni. decisioni ed emozioni*. Bologna: Il Mulino.
Bertolino, Marta. 2013. Imputabilità: scienze, neuroscienze e diritto penale. In *Il diritto nelle neuroscienze. non "siamo" i nostri cervelli*, ed. Laura Palazzani and Roberto Zanotti. Torino: Giappichelli.
Blasi, Gary. 2002. Advocacy Against the Stereotype: Lessons from Cognitive Social Psychology. *UCLA Law Review* 49: 1241–1281.
Blasi, Gary, and John T. Jost. 2006. System Justification Theory and Research: Implications for Law, Legal Advocacy, and Social Justice. *California Law Review* 94: 1119–1168.
Bloom, Paul. 2016. *Against Empathy. The Case for Rational Compassion*. New York: HarperCollins.
Blumenthal, Jeremy A. 2005. Does Mood Influence Moral Judgment? An Empirical Test with Legal and Policy Implications. *Law and Psychology Review* 29: 23.
Blumoff, Theodore Y. 2009. *The Brain Sciences and Criminal Law Norms*. Vol. 1787. Gruter Institute for Law and Behavioral Research Conferences.
Bobbio, Norberto. 1951. La certezza del diritto è un mito? *Rivista internazionale di filosofia del diritto* 28: 151.
Boella, Laura. 2006. *Sentire l'altro. Conoscere e praticare l'empatia*. Milano: Raffaello Cortina.
———. 2011. La morale e la natura. In *Neuroetica. Scienze del cervello, filosofia e libero arbitrio*, ed. Andrea Lavazza and Giuseppe Sartori. Bologna: Il Mulino.
Bona, Carlo. 2010. *Sentenze imperfette. Gli errori cognitivi nei giudizi civili*. Bologna: Il Mulino.
Bona, Carlo, and Rino Rumiati. 2013. *Psicologia cognitiva per il diritto. Ricordare, pensare, decidere nell'esperienza forense*. Bologna: Il Mulino.
Boudon, Raymond. 2009. *Effets pervers et ordre sociale*. Paris: Quadrige - PUF.
Bråten, Stein, ed. 2007. *On Being Moved. From Mirror Neurons to Empathy*. Amsterdam, Philadelphia: John Benjamins.
Burns, Jeffrey M., and Russell H. Swerdlow. 2003. Right Orbitofrontal Tumor with Pedophilia Symptom and Constructional Apraxia Sign. *Archives of Neurology* 60 (3): 437–440.
Callegari, Alessandra. 2017. *Il giudice fra emozioni, biases ed empatia*. Canterano: Aracne.

Calvet Christian, Roberta, and James Alm. 2012. Empathy, Sympathy, and Tax Compliance. *Journal of Economic Psychology* 40: 62–82.

Capraro, Laura. 2013. Primi casi 'clinici' in tema di prova neuroscientifica. In *Il diritto nelle neuroscienze. Non "siamo" i nostri cervelli*, ed. Laura Palazzani and Roberto Zanotti. Torino: Giappichelli.

Cary, Phillip. 2007. A Brief History of the Concept of Free Will: Issues That Are and Are Not Germane to Legal Reasoning. *Behavioral Sciences and the Law* 25: 181.

Castignone, Silvana. 1995. *Diritto, linguaggio, realtà: saggi sul realismo giuridico*. Torino: Giappichelli.

Catellani, Patrizia. 1992. *Il giudice esperto. Psicologia cognitiva e ragionamento giudiziario*. Bologna: Il Mulino.

Caterina, Raffaele, ed. 2008a. *I fondamenti cognitivi del diritto. Percezioni, rappresentazioni, comportamenti*. Torino: Mondadori.

———. 2008b. Un approccio cognitivo alla diversità culturale. In *I fondamenti cognitivi del diritto*, ed. Raffaele Caterina. Torino: Mondadori.

Cesiano, Davide. 2013. L'applicazione Della 'business judgement rule' nella giurisprudenza italiana. *Giurisprudenza Commerciale*, no. 6.

Charness, Gary, and Uri Gneezy. 2008. What's in a Name? Anonymity and Social Distance in Dictator and Ultimatum Games. *Journal of Economic Behavior & Organization* 68 (1): 29–35.

Cicourel, Aaron V. 1974. *Cognitive Sociology. Language and Meaning in Social Interaction*. New York: The Free Press.

Cohen, David, and Jack L. Knetsch. 2000. Judicial Choice and Disparities between Measures of Economic Values. In *Choices, Values, and Frames*, ed. Daniel Kahneman and Amos Tversky. Cambridge: Cambridge University Press.

Cohen, Michael D., James G. March, and Johan P. Olsen. 1972. A Garbage Can Model of Organizational Choice. *Administrative Science Quarterly* 17 (1): 1–25.

Collins, Randall. 1975. *Conflict Sociology. Toward an Explanatory Science*. New York, San Francisco, London: Academic Press.

———. 1987. Interaction Ritual Chains, Power and Property: The Micro-Macro Connection as an Empirically Based Theoretical Problem. In *The Micro-Macro Link*, ed. Jeffrey C. Alexander, Bernhard Giesen, Richard Munch, and Neil J. Smelser. Berkeley, Los Angeles, London: University of California Press.

Crespi, Franco. 1989. *Azione sociale e potere*. Bologna: Il Mulino.

Crozier, Michel. 2010. *The Bureaucratic Phenomenon*. New Brunswick, NJ, London: Transaction Publishers.

D'Avack, Lorenzo. 2013. Neuroscienze ed esperimenti sull'uomo: a partire dall'analisi del parere del comitato nazionale per la bioetica. In *Il diritto nelle neuroscienze. Non "siamo" i nostri cervelli*, ed. Laura Palazzani and Roberto Zanotti. Torino: Giappichelli.

Damasio, Antonio R. 1994. *Descartes' Error. Emotion, Reason, and the Human Brain*. New York: Putnam's and Sons.

Danziger, Shai, Jonathan Levavb, and Liora Avnaim-Pessoa. 2011. Extraneous Factors in Judicial Decisions. *Proceedings of the National Academy of Sciences of the United States of America* 108 (17): 6889–6892.

Dawes, Christopher T., Peter John Loewen, Darren Schreiber, Alan N. Simmons, Taru Flagan, Richard McElreath, Scott E. Bokemper, James H. Fowler, and Martin P. Paulus. 2012. Neural Basis of Egalitarian Behavior. *Proceedings of the National Academy of Sciences of the United States of America* 109 (17): 6479–6483.

Dawes, Robyn M., and John M. Orbell. 1995. The Benefit of Optional Play in Anonymous One-Shot Prisoner's Dilemma Games. In *Barriers to Conflict Resolution*, ed. Kenneth Arrow, Robert H. Mnookin, Lee Ross, Amos Tversky, and Robert Wilson. New York, London: W.W. Norton & Company.

De Caro, Mario. 2011. Libero arbitrio e neuroscienze. In *Neuroetica. Scienze del cervello, filosofia e libero arbitrio*, ed. Andrea Lavazza and Giuseppe Sartori. Bologna: Il Mulino.

de Cataldo Neuburger, Luisella. 1988. *Psicologia della testimonianza e prova testimoniale*. Milano: Giuffrè.

———. 2009. Neuroscienze e diritto penale. La scienza come, quando e perché. In *Le neuroscienze e il diritto*, ed. Amedeo Santosuosso. Como, Pavia: Ibis.

De Felice, Deborah. 2010. Modelli evolutivi e teoria generale del comportamento deviante. In *Neodarwinismo e scienze sociali*, ed. Roberto Vignera. Milano: FrancoAngeli.

———. 2014. *Il sapere specialistico nel giudizio penale. Profili d'analisi di sociologia del diritto*. Milano, Udine: Mimesis.

Den Hartogh, Govert. 2013. The Political Obligation to Donate Organs. *Ratio Juris* 26 (3): 378–403.

Desmoulin-Canselier, Sonia. 2017. Another Perspective on 'Neurolaw': The Use of Brain Imaging in Civil Litigation Regarding Mental Competence. *BioLaw Journal - Rivista di biodiritto* 3 (November): 233–249.

Dhami, Mandeep K. 1982. Legal Decision Making: Psychological Reality Meets Legal Idealism. In *Law and Psychology*, ed. Belinda Brooks-Gordon and Michael Freeman. Oxford: Oxford University Press.

Di Giovine, Ombretta. 2009. *Un diritto penale empatico? Diritto penale, bioetica e neuroetica*. Torino: Giappichelli.

Elster, Jon. 1985. *Sour Grapes: Studies in the Subversion of Rationality*. Cambridge: Cambridge University Press.

Englich, Birte, and Thomas Mussweiler. 2001. Sentencing Under Uncertainty: Anchoring Effects in the Courtroom. *Journal of Applied Social Psychology* 31 (7): 1535–1551.

Farahany, Nita A. 2012. A Neurological Foundation for Freedom. *Stanford Technology Law Review* 4: 1–15.

Farisco, Michele. 2013. Neuroscienze e diritto di cura di pazienti con disturbi della coscienza. In *Il Diritto nelle neuroscienze. Non "siamo" i nostri cervelli*, ed. Laura Palazzani and Roberto Zanotti. Torino: Giappichelli.

Feigenson, Neal. 2006. Brain Imaging and Courtroom Evidence: On the Admissibility and Persuasiveness of fMRI. *International Journal of Law in Context* 2 (3): 233.

Feldman, Yuval, Amos Schurr, and Doron Teichman. 2016. Anchoring Legal Standards. *Journal of Empirical Legal Studies* 13 (2): 298–329.

Ferrari, Pier Francesco, and Vittorio Gallese. 2007. Mirror Neurons and Intersubjectivity. In *On Being Moved. From Mirror Neurons to Empathy*, ed. Stein Bråten. Amsterdam, Philadelphia: John Benjamins.

Festinger, Leon. 1957. *A Theory of Cognitive Dissonance*. Stanford, CA: Stanford University Press.

Finnbogason, Guðmundur. 1913. *L'intelligence sympathique*. Paris: Alcan.

Fisher, Roger, and William L. Ury. 1981. *Getting to Yes. Negotiating Agreement Without Giving In*. New York: Penguin.

Fodor, Jerry A. 1983. *The Modularity of Mind*. Cambridge, MA, London: The MIT Press.

Forza, Antonio. 2010. Neuroscienze e futuri scenari per il diritto. In *Mente, società e diritto*, ed. Guglielmo Gulotta and Antonietta Curci. Milano: Giuffrè.

Forza, Antonio, Giulia Menegon, and Rino Rumiati. 2017. *Il giudice emotivo. La decisione tra ragione ed emozione*. Bologna: Il Mulino.

Frank, Jerome. 1949. *Law and the Modern Mind*. London: Stevens & Sons Limited.

Franks, David D. 2010. *Neurosociology. The Nexus between Neuroscience and Social Psychology*. New York: Springer.

Frey, Bruno S., and Jana Gallus. 2016. Beneficial and Exploitative Nudges. In *Nudging - Possibilities, Limitations and Applications in European Law and Economics*, ed. Klaus Mathis and Avishalom Tor. Cham, Heidelberg, New York, Dordrecht, London: Springer.

Fuselli, Stefano. 2013. Le emozioni nell'esperienza giuridica. L'impatto delle neuroscienze. In *Il diritto nelle neuroscienze. Non "siamo" i nostri cervelli*, ed. Laura Palazzani and Roberto Zanotti. Torino: Giappichelli.

———. 2014. *Diritto, neuroscienze, filosofia: un itinerario*. Milano: FrancoAngeli.
Gallacher, Ian. 2011. Thinking Like Non-lawyers: Why Empathy Is a Core Lawyering Skill and Why Legal Education Should Change to Reflect Its Importance. *Legal Communication & Rhetoric: JALWD* 6: 116.
Garland, Brent, ed. 2004a. *Neuroscience and the Law: Brain, Mind, and the Scales of Justice*. New York, Washington: Dana Press.
———. 2004b. Neuroscience and the Law: A Report. In *Neuroscience and the Law: Brain, Mind, and the Scales of Justice*, ed. Brent Garland. New York, Washington: Dana Press.
Gazzaniga, Michael S. 1985. *The Social Brain. Discovering the Networks of the Mind*. New York: Basic Books.
Genevès, Victor. 2016. Neurosciences et société. Quelle régulation pour quel phénomène? *Lex Electronica*, no. 21. Centre de recherche en droit public, Université de Montréal: 131–147. http://www.lex-electronica.org/articles/volume-21/neurosciences-et-societe-quelle-regulation-pour-quel-phenomene/.
Ghezzi, Morris Lorenzo. 2016. *Il Diritto come estetica. Epistemologia della conoscenza e della volontà: il nichilismo/nihilismo del dubbio*. Milano: Mimesis.
Giddens, Anthony. 1998. *The Third Way. The Renewal of Social Democracy*. Cambridge: Polity Press.
Gigerenzer, Gerd. 2015. On the Supposed Evidence for Libertarian Paternalism. *Review of Philosophy and Psychology* 6 (3): 361–383.
Gintis, Herbert. 2007. A Framework for the Unification of the Behavioral Sciences. *Behavioral and Brain Sciences* 30 (1): 1–16.
Giusberti, Fiorella, Raffaella Nori, Elisa Gambetti, and Luca Bensi. 2010. Emettere una sentenza. Aspetti del processo decisionale. In *Mente, società e diritto*, ed. Guglielmo Gulotta and Antonietta Curci. Milano: Giuffrè.
Glannon, Walter. 2014. The Limitations and Potential of Neuroimaging in the Criminal Law. *The Journal of Ethics* 18 (2). https://doi.org/10.1007/s10892-014-9169-y.
Godsey, Mark. 2017. *Blind Injustice. A Former Prosecutor Exposes the Psychology and Politics of Wrongful Convictions*. Berkeley: University of California Press.
Goldman, David. 2012. *Our Genes, Our Choices. How Genotype and Gene Interactions Affect Behavior*. Amsterdam: Academic Press.
Goodenough, Oliver R. 2006. Can Cognitive Neuroscience Make Psychology a Foundational Discipline for the Study of Law? In *Law and Psychology*, ed. Belinda Brooks-Gordon and Michael Freeman. Oxford: Oxford University Press.
———. 2009. Neuroscientific Developments as a Legal Challenge. In *Le Neuroscienze E Il Diritto*, ed. Amedeo Santosuosso. Como, Pavia: Ibis.
Goodenough, Oliver R., and Kristin Prehn. 2006. A Neuroscientific Approach to Normative Judgment in Law and Justice. In *Law and the Brain*, ed. Semir Zeki and Oliver R. Goodenough. Oxford: Oxford University Press.
Goodenough, Oliver R., and Micaela Tucker. 2010. Law and Cognitive Neuroscience. *Annual Review of Law and Social Science* 6 (1): 61–92.
Greely, Henry T. 2004. Prediction, Litigation, Privacy, and Property. In *Neuroscience and the Law: Brain, Mind, and the Scales of Justice*, ed. Brent Garland. New York: Dana Press.
Green, Joshua, and Jonathan Cohen. 2006. For the Law, Neuroscience Changes Nothing and Everything. In *Law and the Brain*, ed. Semir Zeki and Oliver R. Goodenough. Oxford: Oxford University Press.
Greenberg, Jerald, and Ronald L. Cohen. 1982. Why Justice? Normative and Instrumental Interpretations. In *Equity and Justice in Social Behavior*, ed. Jerald Greenberg and Ronald L. Cohen. New Haven, London: Academic Press.
Greenwald, Anthony G., and Linda Hamilton. 2006. Implicit Bias: Scientific Foundations. *California Law Review* 94: 945–967.
Gregory, Stanford W. 1983. A Quantitative Analysis of Temporal Symmetry in Microsocial Relations. *American Sociological Review* 48 (1): 129–135.

Guala, Francesco, and Luigi Mittone. 2015. A Political Justification of Nudging. *Review of Philosophy and Psychology* 6 (3): 385–395.

Guthrie, Chris. 1999. Better Settle Than Sorry: The Regret Aversion Theory of Litigation Behavior. *University of Illinois Law Review* 47: 43.

Guthrie, Chris, Jeffrey J. Rachlinski, and Andrew J. Wistrich. 2001. Inside the Judicial Mind. *Cornell Law Review* 86 (4): 777–830.

Hagman, William, David Andersson, Daniel Västfjäll, and Gustav Tinghög. 2015. Public Views on Policies Involving Nudges. *Review of Philosophy and Psychology* 6 (3): 439–453.

Hari, Riitta. 2007. Human Mirroring Systems. On Assessing Mind by Reading Brain and Body During Social Interaction. In *On Being Moved. From Mirror Neurons to Empathy*, ed. Stein Bråten. Amsterdam, Philadelphia: John Benjamins.

Haselton, Martie G., and Daniel Nettle. 2006. The Paranoid Optimist: An Integrative Evolutionary Model of Cognitive Biases. *Personality and Social Psychology Review* 10 (1): 47–66.

Hausman, Daniel M., and Brynn Welch. 2010. Debate: To Nudge or Not to Nudge*. *Journal of Political Philosophy* 18 (1): 123–136.

Heller, Kevin Jon. 2006. The Cognitive Psychology of Circumstantial Evidence. *Michigan Law Review* 105: 241–306.

Helleringer, Geneviève. 2016. Designing Disclosures. Testing the Efficacy of Disclosure in Retail Investment Advice. In *Nudging - Possibilities, Limitations and Applications in European Law and Economics*, ed. Klaus Mathis and Avishalom Tor. Cham, Heidelberg, New York, Dordrecht, London: Springer.

Hickok, Gregory. 2009. Eight Problems for the Mirror Neuron Theory of Action Understanding in Monkeys and Humans. *Journal of Cognitive Neuroscience* 21 (7): 1229–1243.

———. 2015. *Il mito dei neuroni specchio. Comunicazione e facoltà cognitive. La nuova frontiera*. Torino: Bollati Boringhieri.

Hillman, Robert A. 2000. The Limits of Behavioral Decision Theory in Legal Analysis: The Case of Liquidated Damages. *Cornell Law Review* 85: 717–738.

Hitlin, Steven, and Stephen Vaisey. 2013. The New Sociology of Morality. *Annual Review of Sociology* 39 (1): 51–68.

Hume, David. 1739. In *A Treatise of Human Nature. [1896]*, ed. L.A. Selby-Bigge. Oxford: Clarendon Press https://www.files.ethz.ch/isn/125487/5010_Hume_Treatise_Human_Nature.pdf.

Iacoboni, Marco. 2009. The Problem of Other Minds Is Not a Problem: Mirror Neurons and Intersubjectivity. In *Mirror Neuron Systems. The Role of Mirroring Processes in Social Cognition*, ed. Jaime A. Pineda. New York: Springer.

Iagulli, Paolo. 2011. *La sociologia delle emozioni*. Milano: FrancoAngeli.

Jhering, Rudolf von. 1913. In *Law as a Means to An End*, ed. Isaac Husik. Boston: The Boston Book Company.

Jolls, Christine, and Cass R. Sunstein. 2004. *Debiasing through Law*. Vol. 225. John M. Olin Law & Economics Working Paper.

———. 2006. The Law of Implicit Bias. *California Law Review* 94: 969–996.

Jones, Owen D., and Sarah F. Brosnan. 2008. Law, Biology, and Property: A New Theory of the Endowment Effect. *William and Mary Law Review* 49 (6): 1935–1990.

Jones, Owen D., Robert Ellickson, Ira Ellman, Adam Gifford, Oliver Goodenough, Lydia Jones, Dennis Karjala, et al. 2001. Time-Shifted Rationality and the Law of Law's Leverage: Behavioral Economics Meets Behavioral Biology. *Northwestern University Law Review* 95 (4): 1141–1206.

Jones, Owen D., and Timothy H. Goldsmith. 2005. Law and Behavioral Biology. *Columbia Law Review* 105: 405–502.

Jones, Owen D., Jeffrey D. Schall, and Francis X. Shen, eds. 2014. *Law and Neuroscience*. New York: Wolters Kluwer.

Jones, Owen D., Anthony D. Wagner, David L. Faigman, and Marcus E. Raichle. 2013. Neuroscientists in Court. *Nature Reviews. Neuroscience* 14 (10): 730–736.

Kahneman, Daniel. 2011. *Thinking, Fast and Slow*. New York: Farrar, Straus and Giroux.
Kahneman, Daniel, and Amos Tversky. 1974. Judgment under Uncertainty: Heuristics and Biases. *Science* 185 (4157): 1124–1131.
———. 2000a. Choices, Values, and Frames. In *Choices, Values, and Frames*, ed. Daniel Kahneman and Amos Tversky. Cambridge: Cambridge University Press.
———. 2000b. Conflict Resolution. A Cognitive Perspective. In *Choices, Values and Frames*, ed. Daniel Kahneman and Amos Tversky. Cambridge, UK: Cambridge University Press.
Kamin, Kim A., and Jeffrey J. Rachlinski. 1995. Ex Post ≠ Ex Ante: Determining Liability in Hindsight. *Law and Human Behavior* 19 (1): 89–104.
Kawohl, Wolfram, and Elmar Habermeyer. 2007. Free Will: Reconciling German Civil Law with Libet's Neurophysiological Studies on the Readiness Potential. *Behavioral Sciences and the Law* 25: 309–320.
Kendon, Adam. 1970. Movement Coordination in Social Interaction: Some Examples Described. *Acta Psychologica* 32 (January): 101–125.
Keysers, Christian, and Valeria Gazzola. 2009. Unifying Social Cognition. In *Mirror Neuron Systems. The Role of Mirroring Processes in Social Cognition*, ed. Jaime A. Pineda. New York: Springer.
Khoshbin, Laura Stephens, and Shahram Khoshbin. 2007. Imaging the Mind, Minding the Image: An Historical Introduction to Brain Imaging and the Law. *American Journal of Law & Medicine* 33: 171–192.
King, Willard L. 1937. The Breakfast Theory of Jurisprudence. *Dicta* 14 (6 (April)): 143–146.
Kittay, Leo. 2007. Admissibility of fMRI Lie Detection. The Cultural Bias Against 'Mind Reading' Devices. *Brooklyn Law Review* 72: 1351–1399.
Kroeber, Hans-Ludwig. 2007. The Historical Debate on Brain and Legal Responsibility — Revisited. *Behavioral Sciences and the Law* 25: 251–261.
Krynski, Tevye R., and Joshua B. Tenenbaum. 2007. The Role of Causality in Judgment under Uncertainty. *Journal of Experimental Psychology. General* 136 (3): 430–450.
Lane, Kristin A., Jerry Kang, and Mahzarin R. Banaji. 2007. Implicit Social Cognition and Law. *Annual Review of Law and Social Science* 3 (1): 427–451.
Langevoorrt, Donald C. 2015. Resetting the Corporate Thermostat: Lessons from the Recent Financial Scandals About Self-Deception, Deceiving Others and the Design of Internal Controls. *Georgetown Law Journal* 93: 285–318.
Lecouteux, Guilhem. 2015. In Search of Lost Nudges. *Review of Philosophy and Psychology* 6 (3): 397–408.
Lepenies, Robert, and Magdalena Małecka. 2015. The Institutional Consequences of Nudging – Nudges, Politics, and the Law. *Review of Philosophy and Psychology* 6 (3): 427–437.
Lerner, Melvin J. 1980. *The Belief in a Just World. A Fundamental Delusion*. New York: Plenum Press.
Levy, Neil. 2014. Is Neurolaw Conceptually Confused? *The Journal of Ethics* 18 (2): 171–185.
Libet, Benjamin, Curtis A. Gleason, Elwood W. Wright, and Dennis K. Pearl. 1983. Time of Conscious Intention to Act in Relation to Onset of Cerebral Activity (Readiness-Potential) - The Unconscious Initiation of a Freely Voluntary Act. *Brain* 106: 623–642.
Lippert-Rasmussen, Kasper. 2014. Neuroprediction, Truth-Sensitivity, and the Law. *The Journal of Ethics* 18 (2): 123–136.
Lodge, Martin, and Kai Wegrich. 2016. The Rationality Paradox of Nudge: Rational Tools of Government in a World of Bounded Rationality. *Law & Policy* 38 (3): 250–267.
Lyons, Derek E. 2009. The Rational Continuum of Human Imitation. In *Mirror Neuron Systems. The Role of Mirroring Processes in Social Cognition*, ed. Jaime A. Pineda. New York: Springer.
Macioce, Fabio. 2013. Le neuroscienze e il processo penale. Una relazione difficile. In *Il diritto nelle neuroscienze. Non "siamo" i nostri cervelli*, ed. Laura Palazzani and Roberto Zanotti. Torino: Giappichelli.
Magrin, Maria Elena. 1995. Psicologia della decisione giudiziaria. In *Manuale di psicologia giuridica*, ed. Assunto Quadrio and Gaetano De Leo. Milano: Edizioni universitarie di lettere economia diritto.

Malouff, John, and Nicola S. Schutte. 1989. Shaping Juror Attitudes: Effects of Requesting Different Damage Amounts in Personal Injury Trials. *The Journal of Social Psychology* 129 (4): 491–497.

Manzotti, Riccardo, and Vincenzo Tagliasco. 2008. *L'esperienza. Perché i neuroni non spiegano tutto*. Torino: Codice Edizioni.

Marchetti, Carlo. 2008. Risparmio, fiducia e corporate governance. In *I fondamenti cognitivi del diritto*, ed. Raffaele Caterina. Torino: Mondadori.

Marrone, Gianfranco. 2011. *Addio alla natura*. Torino: Giulio Einaudi.

Mathis, Klaus, and Avishalom Tor, eds. 2016. *Nudging - Possibilities, Limitations and Applications in European Law and Economics*. Cham: Springer.

Mathis, Klaus, and Philipp Anton Burri. 2016. Nudging in Swiss Contract Law? An Analysis of Non-mandatory Default Rules from a Legal, Economic and Behavioral Perspective. In *Nudging - Possibilities, Limitations and Applications in European Law and Economics*, ed. Klaus Mathis and Avashalom Tor. Cham, Heidelberg, New York, Dordrecht, London: Springer.

Mazzocco, Ketti. 2008. Emozioni e decisione. In *Psicologia del giudizio e della decisione*, ed. Nicolao Bonini, Fabio Del Missier, and Rino Rumiati. Bologna: Il Mulino.

Mead, George Herbert. 1972. In *Mind, Self & Society: From the Standpoint of a Social Behaviorist*, ed. Charles W. Morris. Chicago: University of Chicago Press.

Merton, Robert King. 1968. *Social Theory and Social Structure*. New York: The Free Press.

Messina, Giulia. 2015. I Nuovi orizzonti della prova (neuro)scientifica nel giudizio sull'imputabilità. *Rivista italiana di medicina legale* (1).

Metting, Zwany, Lars A. Rödiger, Jacques De Keyser, and Joukje van der Naalt. 2007. Structural and Functional Neuroimaging in Mild-to-Moderate Head Injury. *The Lancet. Neurology* 6 (8): 699–710.

Miller, George A. 1956. The Magical Number Seven, Plus or Minus Two: Some Limits on Our Capacity for Processing Information. *Psycological Review* 63: 81–97.

Mills, Chris. 2015. The Heteronomy of Choice Architecture. *Review of Philosophy and Psychology* 6 (3): 495–509.

Mitchell, Gregory. 2002. Taking Behavioralism Too Seriously? The Unwarranted Pessimism of the New Behavioral Analysis of Law. *William and Mary Law Review* 43: 1907–2021.

———. 2015. Libertarian Paternalism Is An Oxymoron. *Northwestern University Law Review* 99.

Mnookin, Robert H., and Lee Ross. 1995. Introduction. In *Barriers to Conflict Resolution*, ed. Kenneth Arrow, Robert H. Mnookin, Lee Ross, Amos Tversky, and Robert Wilson. New York, London: W.W. Norton & Company.

Monti, Alberto. 2008. Psicologia della decisione e tutela del consumatore. In *I fondamenti cognitivi del diritto*, ed. Raffaele Caterina. Torino: Mondadori.

Morse, Stephen J. 2004. New Neuroscience, Old Problems. In *Neuroscience and the Law: Brain, Mind, and the Scales of Justice*, ed. Brent Garland. New York: Dana Press.

———. 2009. The Neuroscience Challenges to Criminal Responsibility. In *Le neuroscienze e il diritto*, ed. Amedeo Santosuosso. Como, Pavia: Ibis.

Musatti, Cesare Luigi. 1931. *Elementi di psicologia della testimonianza*. Padova: Cedam.

Nagel, Thomas. 1974. What Is It Like to Be a Bat? *The Philosophical Review* 83 (4): 435–450.

Navarro, Anton D., and Edmund Fantino. 2009. The Sunk-Time Effect: An Exploration. *Journal of Behavioral Decision Making* 22 (3): 252–270.

Niskanen, William A. 1971. *Bureaucracy and Representative Government*. Chicago, IL: Aldine-Atherton.

Norscia, Ivan, and Elisabetta Palagi. 2011. Yawn Contagion and Empathy in Homo Sapiens. *PloS One* 6 (12): e28472.

Nowak, Martin A., Karen M. Page, and K. Sigmund. 2000. Fairness Versus Reason in the Ultimatum Game. *Science* 289 (September): 1773–1775.

Nussbaum, Martha C. 2006. Reply to Amnon Reichman. *Journal of Legal Education* 56: 320–329.

O'Gorman, Rick, David Sloan Wilson, and Ralph R. Miller. 2008. An Evolved Cognitive Bias for Social Norms. *Evolution and Human Behavior* 29 (2): 71–78.

Oberman, Lindsay M., and V.S. Ramachandran. 2009. Reflections on the Mirror Neuron System: Their Evolutionary Functions Beyond Motor Representation. In *Mirror Neuron Systems. The Role of Mirroring Processes in Social Cognition*, ed. Jaime A. Pineda. New York: Springer.
Oliverio, Alberto. 2012. *Cervello*. Torino: Bollati Boringhieri.
Palazzani, Laura. 2013. Potenziamento neuro-cognitivo: aspetti bioetici e biogiuridici. In *Il diritto nelle neuroscienze. Non "siamo" i nostri cervelli*, ed. Laura Palazzani and Roberto Zanotti. Torino: Giappichelli.
Palazzani, Laura, and Roberto Zanotti. 2013. Introduzione. In *Il Diritto Nelle Neuroscienze. Non "Siamo" i Nostri Cervelli*, ed. Laura Palazzani and Roberto Zanotti. Torino: Giappichelli.
Pardo, Michael S., and Dennis Patterson. 2013. *Minds, Brains, and Law. The Conceptual Foundations of Law and Neuroscience*. Oxford: Oxford University Press.
Pattaro, Enrico. 2005. The Law and the Right. A Reappraisal of the Reality That Ought to Be. In *A Treatise*, ed. Enrico Pattaro. Dordrecht: Springer.
Patton, Bruce. 2005. Negotiation. In *The Handbook of Dispute Resolution*, ed. Michael L. Moffit and Robert C. Bordone. San Francisco: Jossey-Bass.
Pettit, Mark, Jr. 2007. FMRI and BF Meet FRE: Brain Imaging and the Federal Rules of Evidence. *American Journal of Law and Medicine* 33: 319–340.
Picozza, Eugenio. 2013. Neurodiritto. Ipotesi di voce per un'enciclopedia o dizionario giuridico. In *Il diritto nelle neuroscienze. Non "siamo" i nostri cervelli*, ed. Laura Palazzani and Roberto Zanotti. Torino: Giappichelli.
Picozza, Eugenio, and Valentina D'Arrigo. 2015. Neurodiritto: nuova filosofia del diritto o teoria generale? In *Diritto, neuroscienze, scienze della cognizione*, ed. Ernesto Fabiani, Sebastiano Faro, and Nicola Lettieri. Napoli: Edizioni scientifiche italiane.
Pietroni, Davide, and Rino Rumiati. 2008. Negoziazione e decisione. In *Psicologia del giudizio e della decisione*, ed. Nicolao Bonini, Fabio Del Missier, and Rino Rumiati. Bologna: Il Mulino.
Pineda, Jaime A. 2009. Preface. In *Mirror Neuron Systems. The Role of Mirroring Processes in Social Cognition*, ed. Jaime A. Pineda. New York: Springer.
Pinker, Steven. 2002. *The Blank Slate. The Modern Denial of Human Nature*. New York: Penguin Books.
Pizzetti, Federico Gustavo. 2012. *Neuroscienze Forensi e diritti fondamentali: spunti costituzionali*. Torino: Giappichelli.
Pockett, Susan. 2007. The Concept of Free Will: Philosophy, Neuroscience and the Law. *Behavioral Sciences and the Law* 25 (February): 281–293.
Pogliani, Anna. 1989. L'immagine della giustizia in Italia: una ricerca sulle opinioni e sugli atteggiamenti. In *Giustizia, norme e autoregolazione in psicologia sociale*, ed. Fiorangela Oneroso Di Lisa. Napoli: Liguori Editori.
Posner, Richard A. 2004. *Frontiers of Legal Theory*. Cambridge, MA, London: Harvard University Press.
Rachlinski, Jeffrey J. 1982. Cognitive Errors, Individual Differences, and Paternalism. In *Law and Psychology*, ed. Belinda Brooks-Gordon and Michael Freeman. Oxford: Oxford University Press.
———. 1998. A Positive Psychological Theory of Judging in Hindsight. *University of Chicago Law Review* 65: 571–625.
———. 2000a. Heuristics and Biases in the Court: Ignorance or Adaptation? *Oregon Law Review* 79: 61.
———. 2000b. The 'New' Law and Psychology: A Reply to Critics, Skeptics, and Cautious Supporters. *Cornell Law Review* 85: 739–766.
———. 2002. The Uncertain Psychological Case for Paternalism. *Northwestern University Law Review* 97: 1165–1226.
Rachlinski, Jeffrey J., and Forest Jourden. 1998. Remedies and the Psychology of Ownership. *Vanderbilt Law Review* 51: 1541.
Rachlinski, Jeffrey J., and Andrew J. Wistrich. 2017. Judging the Judiciary by the Numbers: Empirical Research on Judges. *Annual Review of Law and Social Science* 13 (1): 203–229.

Ragone, Giada, and Benedetta Vimercati. 2017. Neuroscience and End-of-Life Decisions. New Anthropological Challenges for Constitutional Law: "Is Human Nature the Only Science of Man"? *BioLaw Journal - Rivista di biodiritto* 3 (November): 111–130.

Raine, Adrian. 2014. *The Anatomy of Violence. The Biological Roots of Crime*. New York: Vintage Books.

Rangone, Nicoletta. 2012. Errori cognitivi e scelte di regolazione. *Analisi giuridica dell'economia* 1: 7–18.

Ridley, Matt. 1996. *The Origins of Virtue*. New York: Penguin Books.

Rizzolatti, Giacomo, and Laila Craighero. 2004. The Mirror-Neuron System. *Annual Review of Neuroscience* 27: 169–192.

Rizzolatti, Giacomo, and Corrado Sinigaglia. 2006. *So quel che fai. Il cervello che agisce e i neuroni specchio*. Milano: Raffaello Cortina.

Rochat, Philippe, and Claudia Passos-Ferreira. 2009. From Imitation to Reciprocation and Mutual Recognition. In *Mirror Neuron Systems. The Role of Mirroring Processes in Social Cognition*, ed. Jaime A. Pineda. New York: Springer.

Rose, Nikolas. 2007. *The Politics of Life Itself: Biomedicine, Power, and Subjectivity in the Twenty-First Century*. Princeton, Oxford: Princeton University Press.

Ross, Lee. 1995. Reactive Devaluation in Negotiation and Conflict Resolution. In *Barriers to Conflict Resolution*, ed. Kenneth Arrow, Robert H. Mnookin, Lee Ross, Amos Tversky, and Robert Wilson. New York, London: W.W. Norton & Company.

Rumiati, Rino, and Lorella Lotto. 2011. Decisioni e decisioni morali, tra razionalità ed emozioni. In *Neuroetica. Scienze del cervello, filosofia e libero arbitrio*, ed. Andrea Lavazza and Giuseppe Sartori. Bologna: Il Mulino.

Sammicheli, Luca, and Giuseppe Sartori. 2010. Delitto, geni, follia. In *Neodarwinismo e scienze sociali*, ed. Roberto Vignera. Milano: FrancoAngeli.

Santosuosso, Amedeo. 2009. Il dilemma del diritto di fronte alle neuroscienze. In *Le neuroscienze e il diritto*, ed. Amedeo Santosuosso. Como, Pavia: Ibis.

———. 2011. *Diritto, scienza, nuove tecnologie*. Padova: Cedam.

Sapolsky, Robert M. 2006. The Frontal Cortex and the Criminal Justice System. In *Law and the Brain*, ed. Semir Zeki and Oliver R. Goodenough. Oxford: Oxford University Press.

Sartor, Giovanni. 2005. Legal Reasoning. A Cognitive Approach to the Law. In *A Treatise*, ed. Enrico Pattaro. Dordrecht: Springer.

Sartori, Giuseppe, Andrea Lavazza, and Luca Sammicheli. 2011. Cervello, diritto e giustizia. In *Neuroetica. Scienze del cervello, filosofia e libero arbitrio*, ed. Andrea Lavazza and Giuseppe Sartori. Bologna: Il Mulino.

Sawicki, Nadia N. 2016. Ethical Limitations on the State's Use of Arational Persuasion. *Law & Policy* 38 (3): 211–233.

Schauer, Frederick. 2010. Neuroscience, Lie-Detection, and the Law: Contrary to the Prevailing View, the Suitability of Brain-Based Lie-Detection for Courtroom or Forensic Use Should Be Determined According to Legal and Not Scientific Standards. *Trends in Cognitive Sciences* 14 (3): 101–103.

Schütz, Alfred. 1967. *The Phenomenology of the Social World*. Chicago: Northwestern University Press.

Searle, John R. 1997. *The Mistery of Consciousness*. New York: The NY Review of Books.

Shafi, Noel. 2009. Neuroscience and Law: The Evidentiary Value of Brain Imaging. *Graduate Student Journal of Psychology* 11: 27–39.

Simon, Herbert A. 1957. *Models of Man*. New York, London: John Wiley - Chapman & Hall.

Sood, Avani Mehta. 2013. Motivated Cognition in Legal Judgments. An Analytic Review. *Annual Review of Law and Social Science* 9 (1): 307–325.

Spranger, Tade M. 2009. A German Perspective on Legal Issues Arising from Neuroscientific Research and Application. In *Le Neuroscienze E Il Diritto*, ed. Amedeo Santosuosso. Como, Pavia: Ibis.

Stake, Jeffrey Evans. 2006. The Property 'Instinct'. In *Law and the Brain*, ed. Semir Zeki and Oliver R. Goodenough. Oxford: Oxford University Press.

Steffen, Ariel David. 2016. Judging Is Nudging: The Inevitability of Value Judgments. Consequences of the Collapse of the Fact/Value Dichotomy for the Behavioural Law and Economics. In *Nudging - Possibilities, Limitations and Applications in European Law and Economics*. Cham, Heidelberg, New York, Dordrecht, London: Springer.

Stevens-Simon, Catherine, Jeffrey I. Dolgan, Lisa Kelly, and Dena Singer. 1997. The Effect of Monetary Incentives and Peer Support Groups on Repeat Adolescent Pregnancies. A Randomized Trial of the Dollar-a-Day Program. *JAMA: The Journal of the American Medical Association* 277 (12): 977–982.

Suchman, Mark C. 1997. On Beyond Interest: Rational, Normative and Cognitive Perspectives in the Social Scientific Study of Law. *Wisconsin Law Review* 3: 475.

Sunstein, Cass R. 2000. Deliberative Trouble - Why Groups Go to Extremes. *Yale Law Journal* 110: 71–119.

———. 2014. Choosing Not to Choose. *Duke Law Journal* 64 (1): 1.

———. 2015. Nudges, Agency, and Abstraction: A Reply to Critics. *Review of Philosophy and Psychology* 6 (3): 511–529.

———. 2017. *Human Agency and Behavioral Economics: Nudging Fast and Slow*. London: Palgrave Macmillan.

Tafaro, Laura. 2017. Neuroscienze e diritto civile: nuove prospettive. *BioLaw Journal - Rivista Di biodiritto* 3 (November): 251–272.

Tancredi, Laurence R. 2004. Neuroscience Developments and the Law. In *Neuroscience and the Law: Brain, Mind, and the Scales of Justice*, ed. Brent Garland. New York, Washington: Dana Press.

Tereszkiewicz, Piotr. 2016. Neutral Third-Party Counselling as Nudge Toward Safer Financial Products? The Case of Risky Mortgage Loan Contracts. In *Nudging - Possibilities, Limitations and Applications in European Law and Economics*, ed. Klaus Mathis and Avishalom Tor. Cham, Heidelberg, New York, Dordrecht, London: Springer.

Terracina, David. 2013. Diritto penale e neuroscienze. In *Il diritto nelle neuroscienze. Non "siamo" i nostri cervelli*, ed. Laura Palazzani and Roberto Zanotti. Torino: Giappichelli.

Thaler, Richard H. 1980. Toward a Positive Theory of Consumer Choice. *Journal of Economic Behavior & Organization* 1: 39–60.

Thaler, Richard H., and Cass R. Sunstein. 2009. *Nudge. Improving Decisions About Health, Wealth and Happyness*. London: Penguin Books.

Tisseron, Serge. 2010. *L'empathie au coeur du jeu social*. Paris: Albin Michel.

Tooby, John, and Lela Cosmides. 1992. The Psychological Foundations of Culture. In *The Adapted Mind. Evolutionary Psychology and the Generation of Culture*, ed. Jerome H. Barkow, Leda Cosmides, and John Tooby. New York, Oxford: Oxford University Press.

Trout, J.D. 2005. Paternalism and Cognitive Bias. *Law and Philosophy* 24 (4): 393–434.

van Baaren, Rick B., Rob W. Holland, Bregje Steenaert, and Ad van Knippenberg. 2003. Mimicry for Money: Behavioral Consequences of Imitation. *Journal of Experimental Social Psychology* 39 (4): 393–398.

Watkins, Daniel. 2010. A Nudge to Mediate: How Adjustments in Choice Architecture Can Lead to Better Dispute Resolution Decisions. *The American Journal of Mediation* 4: 10.

Wentworth, William, and Darrell Yardley. 1994. Deep Sociality: A Bioevolutionary Perspective on the Sociology of Emotions. In *Social Perspectives on Emotion*, ed. William M. Wentworth and John Ryan. Bingley: Emerald.

Wexler, Bruce E. 2008. *Brain and Culture. Neurobiology, Ideology, and Social Change*. Cambridge, MA: MIT Press.

Whitman, Douglas Glen, and Mario J. Rizzo. 2015. The Problematic Welfare Standards of Behavioral Paternalism. *Review of Philosophy and Psychology* 6 (3): 409–425.

Wilkinson, T.M. 2013. Nudging and Manipulation. *Political Studies* 61 (2): 341–355.

Williams, Simon. 2001. *Emotion and Social Theory*. Thousand Oaks, London, New Dehli: Sage.

Wright, Robert. 1995. *The Moral Animal: Why We Are The Way We Are*. New York: Vintage.

Chapter 5
Conclusion

In this work I argued that if sociology and the sociology of law do not take the other disciplines seriously into account, they risk losing their scientific utility. As much as we can still feel intellectually in debt to Weber, Simmel, or Durkheim, I share Searle's assessment that their theoretical achievements cannot be useful here and now, because they do not afford an adequate knowledge of intentionality and individual rule-governed behavior. Disciplinary structuring and compartmentalization in academia have played a useful political role, because they have enabled knowledge to become institutionally independent. However, the makers of this disciplinary knowledge should not be fooled into thinking that mastery in their field can point them in the right direction. Therefore, my first hypothesis has been primarily methodological, and it represents the work's *pars destruens*: the social sciences, whenever they remain in isolation, find it increasingly difficult to formulate plausible predictions or explanations, and can only provide partially verifiable theories. Some disciplines have understood this in advance. Economics and psychology, for example, have been less timid, while I see a stronger risk of marginalization for sociology.

In this book, I have set out a common framework under which it may be possible to integrate four major approaches to the study of law-oriented behavior: the sociological approach, the psychological, the biological, and the cognitive. I leave it to the reader to assess whether the attempt has been successful. The discussion has left out some fundamental social disciplines, such as economics and history, while on the side of the natural sciences, in the "biological" chapter, the evolutionary and epigenetic approaches to behavior have perhaps been inappropriately combined. For this reason alone, purists could raise eyebrows. Physics has completely been left out, too, not only for reasons of space and in the interests of balance between the sections of the book, but also in order not to misuse a discipline whose nuances may well escape me.

But this work takes aim precisely at purism, and it is paradoxical that sociology has ended up indulging in methodological purism just to defend its theoretical holism. The cognitive legal sociology I propose here requires methodological

L. Cominelli, *Cognition of the Law*,
https://doi.org/10.1007/978-3-319-89348-8_5

syncretism in order to make sense of diverse and sometimes conflicting empirical results. The individualistic position expounded in the first chapter concerns, as is obvious, not so much the overall methodological approach to the study of behavior as the unit of analysis at which it is necessary to target the sociological approach, and with which the other methods will integrate syncretically. In the latest trend, research has moved in the opposite direction, and in several passages I pointed out the wise but isolated voices that have attributed this tendency, not to a love for the purity of concepts, but to intellectual indolence and institutional convenience. From the physics of subatomic particles to the functioning of constitutional courts, there is a continuity that we cannot continue to deny, if not for the sake of convenience.

It was precisely in the work of the old masters that sociology was conceived as an interdisciplinary science. Today, the original definition of sociology as "social physics" should not seem so incomprehensible. Likewise, it should be clear that Weber's theory of social action is built on solid historical, legal, and economic concepts, and essentially on a system that structurally resembles analytical mechanics. We should not forget this multidisciplinary origin today, especially in the face of similarities or hybridizations that may seem risky or lack a solid theoretical foundation. And yet disciplines that are strictly related within the social sciences offer visions that are inexplicably incompatible with the natural history of the world, and inconsistent with one another, and I do not shrink from characterizing this state of affairs as scientifically appalling. As useful and necessary as it may be to tailor our methods to different tasks, this differentiation cannot warrant the methodological isolation in which our research communities seem to live.

Someone might fear that to revive the interdisciplinary core of the social sciences is to get the "flow of knowledge" to advance in only one direction, from the natural sciences to the social sciences themselves. But the philosophical and social implications of the great theories of natural science, from Darwinian natural selection to quantum mechanics, can elude only those who have observed their mathematical formality in a superficial way. If this requires too difficult an adaptation for the social scientist, it is not for me to say. The sociological tools of enquiry should make it possible to comfortably frequent the paradigms of a *general* social science that is differentiated only in virtue of its means of investigation. This would not necessarily require a mastery of mathematical or statistical tools, but it certainly implies a willingness to question their methodological certainties, and to formulate integrated paradigms for the collection of empirical data.

As a matter of fact, sociology has already signaled in many ways that it sees the benefit of taking an interdisciplinary approach in the study of human nature. One of the cornerstones of the present-day cognitive sociology of law lies in Mead's nonreductionist evolutionary perspective. Mead's analysis has legitimated the notion that social practices can be instinctual, and that they can also be explained as biological evolutionary processes. It is true that the social behaviors Mead refers to are mainly acts of communication, but the overlap of communicative and normative processes suggests that the same principle can be usefully be brought to bear in explaining our behavior as actors moving about within the sphere of the law. In this way, I have sought to preempt charges of pure behaviorism.

In order to avoid misunderstandings, I should clarify some further concepts. There is such a thing as an "epistemological behaviorism" which needs to be understood in a broader sense, and which should not be labelled reductionist. A sociologist who describes behaviors may be loosely defined as a behaviorist. How to define this approach in another way? It may perhaps suffice to recall that neobehaviorism has overcome this kind of criticism by integrating behavior with consciousness and perception, but my impression is that these kinds of objections would persist. For, in the end, one will always be liable to charges of determinism or biological reductionism.

Any science, however, can be described as broadly reductionist if it seeks to arrive at the root causes of some phenomenon. Evolutionary psychology was certainly deterministic insofar as it assumed that every system of rules, and so also of legal rules, responds *solely* to a natural instinct for cooperation that in the long run makes it convenient for living beings to behave well toward others. The key point lies in that "solely." This game of posturing has fortunately been disrupted by the mutual feedback (both demonstrated and demonstrable) between sociality and naturality.

While the empirical stance this entire work is predicated on scarcely needs any more emphasizing, it is well to provide it with some explanations and clarifications. The empirical approach—along with the criticism it directs at sociological idealism and at the supremacy of pure concepts that essentially result in a metaphysic—forms the basis of the cognitive sociology of law. The previously discussed "bridge-laws"—essential in linking up our perceptions of justice, our cognitive biases, and our behaviors within the space of the law—are difficult to reformulate without making reference to measurable factual events. The situation today is obviously different from the opportunities that ethnomethodologists could look to: we are now able to more accurately measure perceptions and emotions.

On the other hand, I have been able to support the point that the "epistemological wall" separating the natural sciences from the human sciences is more brittle than we used to think. Physics and biology, too, find it necessary to refer to bridge-laws in order to link up the paradigms of the hard sciences and to justify their conceptual apparatus in the face of seemingly contradictory, inconstant, and nonreplicable measurements. Hence the epistemological proposal of adopting a weak version of physicalism, with which the first chapter ended, and which has been pursued consistently as the work unfolded.

Another starting point was the evolutionary principle: while this principle may not allow us to fully explain regulatory phenomena, it does offer interesting hypotheses that need to be tested case by case. And since biology can always suggest a plausible reason for most life events, we cannot afford to ignore this line of explanation. Consider, too, that it was in fact biology, with the Darwinian theory, that borrowed the evolutionary principle from philosophy and the social sciences. In this way, the cultural variable cannot be expelled as some would fear, and as long as it does not pretend to represent an absolute truth, it retains a dialectical relationship with the biological variable. The challenge is to demonstrate that human beings are necessarily cultural animals precisely because of their biology, as is any other animal with a transmissible culture: they are "animals" because they have evolved over a timescale that is beyond the reach of the individual (hence their instinctual cognitive capacities), and they are "cultural" because at the individual level, they can influence the behavior of their peers and descendants to a remarkable degree. I am in good company in making the case that the horror of biological determinism or reductionism brings with it the equally lethal risk of sociological determinism. Because we loathe Lombrosianism, we swung over into the opposite excess, believing that humans are a *tabula rasa*, and hence entirely determined by the environment. Although we intuitively know that this cannot be true—because we can see it with our eyes every day, in the form of individual predispositions to certain behaviors or specific illnesses—we have accepted the double standard under which sociological theory can be contradicted by empirical evidence and yet still be held up as valid.

But we cannot, of course, advance theories that are not amenable to testing. The testing I am proposing is socio-cognitive, and the hypothesis I am putting forward for testing is that like any form of behavior, so the behavior through which we relate to the law is guided and framed by *biosocial* constraints that are neither entirely genetic nor entirely cultural. By the prefix *bio-*, however, I mean something different from what is generally meant in the social sciences, something that can be understood only with the discovery of the mutual feedback between genotype and phenotype as described in epigenetic research. *Biosocial* is a familiar label that invites objections, and the association is almost Pavlovian: the moment we speak of biological constraints, the social scientist is reminded of social Darwinism, and we have to deal with that. So, therefore, mindful that not too much can be made to hang on a label, I do think it wise to forestall these objections by proposing that *biosocial* be replaced with *cognitive*, and by reinterpreting this word so as to designate the framework of instincts and constructs within which our learning, information, evaluation and decision-making processes take place, that is, within which our cognition takes place. As I hope to have made clear in the discussion, this focus on cognition is not a way to smuggle in a behaviorist approach in disguise, and there is such a range of approaches to the study of cognition that we need not worry about it conjuring up any single stereotype. At the same time, we should be wary of fashioning cognition into a totalizing paradigm: just as not everything can be resolved into communication or evolution, so not everything can be resolved into cognition.

Along with the cognitive hypothesis, I am putting forward a corollary specifically related to law, namely, that legal systems are pseudo-moral rules deriving from our biosociality, and that for a proper understanding of legal emotions (somewhat different from our "legal consciousness") we need to also draw on the paradigms of cognitive science, among the others needed to form a full picture of our attitudes to the law. Sociologically, then, the law can be seen as a *super-meme*: a biosocial constraint that develops in complex societies. The super-meme hypothesis may explain the spurious contradiction between law as a static and historical phenomenon and law as a dynamic and promotional element.

In discussing the cognitive biases that affect our *legal behaviors*, we previously took up the distinction between *defensive* and *assertive* biases, the former *safety-seeking*, the latter *risk-taking*. It is precisely by balancing these opposite biases that we can make sense of the ambivalence of social behavior, always torn between the comfort of protection and the challenge of the unknown. That this is a false contradiction was already highlighted in Hume's analysis of the sense of justice, which on the one hand arises artificially—the outcome of education and conventions—but at the same time is shaped by human circumstances and needs. But this means that the sense of equity, on which law and justice are based, is defined by principles that may come into conflict: equality, need, retribution, protection, reciprocity, deserts. On a cognitive approach, we would be able to underscore that these principles correspond to cognitive impulses and mechanisms that distort our perceptions depending on context.

Popular wisdom is replete with proverbs and rules of conduct that are similarly conflicting. As was argued earlier on, it is only to some extent that this can be considered incoherent. This evolutionarily minded vision would definitively dispel the Rousseauvian myth of humanity's loss of innocence, assuming we still need that service, by painting an image of all living beings as inherently individualistic yet strongly inclined toward cooperation. We are naturally inclined to impose general and binding rules just to guard our autonomy, and we are particularly sensitive to violations of these rules. At a rapid pace, this dynamic is increasingly taking hold from generation to generation according to a logic that displaces evolutionary mechanisms in the strict sense.

I did not ignore the various criticisms that have been directed at the cognitive theory of law and in general at behavioral theories of decision-making. Some of these criticisms may be addressed by pointing out that the cognitive theory of law still considers it to be its task to fully understand in what contexts and under what conditions our biases come into play. By investigating our cognitive mechanisms, we can gain a deeper understanding of the motives behind the behavior that we and others engage in within the scope of the law. The law—acting in our own interest, if need be against our will—obligates us to insure ourselves against various contingencies (such as car accidents and bad health), or it compels us to set aside resources for old age. As much as these measures may seem to be a point of arrival, the study of cognitive biases shows that we may push the idea even further. Will it one day be possible to lower our insurance premium by taking a psychological test that measures our aversion to risk? Insurance companies are starting to adjust car insurance premiums to our driving habits, if we consent to have them telemetrically measured: thus, if it turns out that we tend to respect the speed limit, this will lower our premium; conversely, if it turns out that we have a propensity to speed, our premium will be raised (regardless of any speeding tickets we may receive).

If we are willing to bind ourselves to a contractual provision of this kind, is this not in itself evidence that we recognize our cognitive limits and the need to protect ourselves accordingly? If I account myself to be a diligent person but fear that I may be liable to act in a negligent way under stress (as by speeding when irritable), then it makes sense for me to forestall that possibility by signing away my own control over my attitude to risk. Nor does it now seem to be such a remote possibility for contracting parties to submit to cognitive profiling. Should this be considered discriminatory, like denying health insurance for preexisting conditions tied to a genetic risk factor? And, if so, should we not also reject, by the same principle, the practice of tying insurance premiums to a policyholder's age or residence, even without knowing anything else about this person (a form of price discrimination the law paradoxically permits)?

Legal systems have also dealt with cognitive biases through disclosure requirements and through obligations to keep the public well informed, but these measures are often implemented naively and carried out counterproductively. Requirements to keep the public informed easily translate into unnecessary cognitive overload for consumers or into formalities that lend themselves to perfunctory execution, as when we find ourselves placing multiple signatures on contract clauses or informed consent forms. With a better understanding of cognitive mechanisms, we would be able to find more effective ways to ensure that the public understands the consumer protections available to it, as well as other protected interests, without flooding people with information they are likely to find empty or burdening them with formalities that feel like a nuisance.

This last example serves to illustrate the need for proper nudging (Sect. 4.7), but it also highlights how complex it is to influence and steer individual behavior, if not toward social utility, at least away from social dysfunctionality. The purpose of the empirical research program advanced in cognitive sociology will be to ascertain whether a cognitively smart law can achieve a balanced asymmetric paternalism, the ideal being to keep risks away from those who are susceptible to tragic mistakes, without placing an unreasonable burden on those who are not. This approach will provide the social sciences with promising research hypotheses about law, rather than going the route of neuroscientific techniques or genetic therapies that promise to intervene on the genotype in a targeted manner so as to prevent aging or even to fix character flaws or anomic and antisocial personalities. It will be only natural for humans to become "artificial," but it would be unwise to speed up this process without allowing it to take its gradual course.

The manufacturer's authorised representative in the EU is Springer Nature Customer Service Centre GmbH, Europaplatz 3, 69115 Heidelberg, Germany. If you have any concerns regarding our products, please contact ProductSafety@springernature.com

Printed and bound by CPI Group (UK) Ltd, Croydon, CR0 4YY

30/09/2026

02225195-0014